Praise for **ON INTELLIGENCE**

"A manifesto for a theory based on an elegantly simple premise: that intelligence is rooted in the brain's ability to access memories rather than in its ability to process new data."
—*BusinessWeek*

"Provocative thoughts on how the brain and the mind may actually function."
—*Scientific American Mind*

"*On Intelligence* will have a big impact; everyone should read it. In the same way that Erwin Schrödinger's 1943 classic *What Is Life?* made how molecules store genetic information the big problem for biology at that time, *On Intelligence* lays out the framework for understanding the brain."
—James D. Watson, chancellor, Cold Spring Harbor Laboratory, and 1962 Nobel laureate in physiology

"Jeff Hawkins' *On Intelligence* is an important book. It lays out a dramatic new unified theory of how the brain works, in a way that even lay readers can easily understand. And it predicts an exciting technology future filled with truly intelligent machines that go far beyond today's computers and crude robots."
—Walter S. Mossberg, columnist, *The Wall Street Journal*

"A must-read for anyone who has an interest in how we think."
—*Handheld Computing*

" 'Pops the hood' of the neocortex and carefully articulates a theory of consciousness and intelligence that offers radical options for future researchers . . . His engaging speculations are sure to win fans of authors like Steven Johnson and Daniel Dennett."
—*Publishers Weekly*

"Read this book. Burn all the others. It is original, inventive, and thoughtful, from one of the world's foremost thinkers. Jeff Hawkins will change the way the world thinks about intelligence and the prospect of intelligent machines."
—John Doerr, partner, Kleiner Perkins Caufield & Byers

"A landmark book. *On Intelligence* is the first clear exposition of what could be the long-awaited 'great general theory' of human brain function. Loaded with intelligence, insight, and wisdom, it's a wonderfully readable account of the fundamental principles of the brain by a great American original."

—Mike Merzenich, professor of neuroscience,
University of California, San Francisco

"Brilliant and embued with startling clarity. *On Intelligence* is the most important book in neuroscience, psychology, and artificial intelligence in a generation."

—Malcolm Young, professor of biology and provost,
University of Newcastle

"A remarkable synthesis of ideas. I predict it will have a major impact on neuroscience and neuroscientists."

—Mriganka Sur, professor of neuroscience,
Massachusetts Institute of Technology

"A brilliantly presented and innovative hypothesis of how the brain works." —Patrick McGovern, founder and chairman, IDG

"Jeff Hawkins has written an original, thought-provoking and, with the help of Sandra Blakeslee, remarkably readable book that presents a new theory of the functions of the cerebral cortex in perception, cognition, action, and intelligence. What is distinctive about his theory is the original way existing ideas about the cerebral cortex and its architecture have been combined and elaborated based on an extensive knowledge of how the brain works—what Hawkins calls Real Intelligence in contrast to computer-based Artificial Intelligence. As a result this book is a must read for everyone who is curious about the brain and wonders how it works. Many sections of this book, especially those on intelligence, creativity, and minds of silicon, are so thoughtful and original that they are likely to be required reading for college undergraduates for years to come."

—Eric R. Kandel, professor, Columbia University,
senior investigator, Howard Hughes Medical Institute,
and 2000 Nobel Laureate in medicine

on
INTELLIGENCE

INTELLIGENCE

JEFF HAWKINS
WITH SANDRA BLAKESLEE

AN OWL BOOK
Henry Holt and Company
New York

Owl Books
Henry Holt and Company, LLC
Publishers since 1866
175 Fifth Avenue
New York, New York 10010
www.henryholt.com

An Owl Book® and ® are registered trademarks of Henry Holt and Company, LLC.

Distributed in Canada by H. B. Fenn and Company Ltd.

Library of Congress Cataloging-in-Publication Data
Hawkins, Jeff, 1957–
On intelligence / Jeff Hawkins with Sandra Blakeslee.—1st ed.
p. cm.
Includes bibliographical references and index.
ISBN-13: 978-0-8050-7853-4
ISBN-10: 0-8050-7853-3
1. Brain. 2. Intellect. 3. Artificial intelligence. 4. Neural networks (Computer science). I. Blakeslee, Sandra. II. Title.
QP376.H294 2004
612.8'2—dc22 2004051640

Henry Holt books are available for special promotions and premiums.
For details contact: Director, Special Markets.

Originally published in hardcover in 2004 by Times Books

First Owl Books Edition 2005

DESIGNED BY RENATO STANISIC

Printed in the United States of America
10 9 8 7 6 5 4 3 2 1

CONTENTS

Prologue 1

1. *Artificial Intelligence 9*
2. *Neural Networks 23*
3. *The Human Brain 40*
4. *Memory 65*
5. *A New Framework of Intelligence 85*
6. *How the Cortex Works 106*
7. *Consciousness and Creativity 177*
8. *The Future of Intelligence 205*

Epilogue 234

Appendix: Testable Predictions 237

Bibliography 247

Acknowledgments 253

Index 255

on
INTELLIGENCE

PROLOGUE

This book and my life are animated by two passions.

For twenty-five years I have been passionate about mobile computing. In the high-tech world of Silicon Valley, I am known for starting two companies, Palm Computing and Handspring, and as the architect of many handheld computers and cell phones such as the PalmPilot and the Treo.

But I have a second passion that predates my interest in computers—one I view as more important. I am crazy about brains. I want to understand how the brain works, not just from a philosophical perspective, not just in a general way, but in a detailed nuts and bolts engineering way. My desire is not only to understand what intelligence is and how the brain works, but how to build machines that work the same way. I want to build truly intelligent machines.

The question of intelligence is the last great terrestrial frontier of science. Most big scientific questions involve the very small, the very large, or events that occurred billions of years ago. But everyone has a brain. You are your brain. If you want to understand why you feel the way you do, how you perceive the world, why you make mistakes, how you are able to be creative, why music and art are inspiring, indeed what it is to be human, then you need to understand the brain. In addition, a successful theory of intelligence and brain function will have large societal benefits, and not just in helping us cure brain-related diseases. We will be able to build genuinely intelligent machines, although they won't be anything like the robots of popular fiction and computer science fantasy. Rather, intelligent machines will arise from a new set of principles about the nature of intelligence. As such, they will help us accelerate our knowledge of the world, help us explore the universe, and make the world safer. And along the way, a large industry will be created.

Fortunately, we live at a time when the problem of understanding intelligence can be solved. Our generation has access to a mountain of data about the brain, collected over hundreds of years, and the rate at which we are gathering more data is accelerating. The United States alone has thousands of neuroscientists. Yet we have no productive theories about what intelligence is or how the brain works as a whole. Most neurobiologists don't think much about overall theories of the brain because they're engrossed in doing experiments to collect more data about the brain's many subsystems. And although legions of computer programmers have tried to make computers intelligent, they have failed. I believe they will continue to fail as long as they keep ignoring the differences between computers and brains.

What then is intelligence such that brains have it but computers don't? Why can a six-year-old hop gracefully from rock to rock in a streambed while the most advanced robots of our time are lumbering zombies? Why are three-year-olds already well on their way to mastering language while computers can't, despite half a century of programmers' best efforts? Why can you tell a cat from a dog in a fraction of a second while a supercomputer cannot make the distinction at all? These are great mysteries waiting for an answer. We have plenty of clues; what we need now are a few critical insights.

You may be wondering why a computer designer is writing a book about brains. Or put another way, if I love brains why didn't I make a career in brain science or in artificial intelligence? The answer is I tried to, several times, but I refused to study the problem of intelligence as others have before me. I believe the best way to solve this problem is to use the detailed biology of the brain as a constraint and as a guide, yet think about intelligence as a computational problem—a position somewhere between biology and computer science. Many biologists tend to reject or ignore the idea of thinking of the brain in computational terms, and computer scientists often don't believe they have anything to learn from biology. Also, the world of science is less accepting of risk than the world of business. In technology businesses, a person who pursues a new idea with a reasoned approach can enhance his or her career regardless of whether the particular idea turns out to be successful. Many successful entrepreneurs achieved success only after earlier failures. But in academia, a couple of years spent pursuing a new idea that does not work out can permanently ruin a young career. So I pursued the two passions in my life simultaneously, believing that success in industry would help me achieve success in understanding the brain. I needed the financial resources to pursue the science I wanted, and I needed to learn how to affect change in the world, how to sell new ideas, all of which I hoped to get from working in Silicon Valley.

In August 2002 I started a research center, the Redwood Neuroscience Institute (RNI), dedicated to brain theory. There are many neuroscience centers in the world, but no others are dedicated to finding an overall theoretical understanding of the neocortex—the part of the human brain responsible for intelligence. That is all we study at RNI. In many ways, RNI is like a start-up company. We are pursuing a dream that some people think is unattainable, but we are lucky to have a great group of people, and our efforts are starting to bear fruit.

The agenda for this book is ambitious. It describes a comprehensive theory of how the brain works. It describes what intelligence is and how your brain creates it. The theory I present is not a completely new one. Many of the individual ideas you are about to read have existed in some form or another before, but not together in a coherent fashion. This should be expected. It is said that "new ideas" are often old ideas repackaged and reinterpreted. That certainly applies to the theory proposed here, but packaging and interpretation can make a world of difference, the difference between a mass of details and a satisfying theory. I hope it strikes you the way it does many people. A typical reaction I hear is, "It makes sense. I wouldn't have thought of intelligence this way, but now that you describe it to me I can see how it all fits together." With this knowledge most people start to see themselves a little differently. You start to observe your own behavior saying, "I understand what just happened in my head." Hopefully when you have finished this book, you will have new insight into why you think what you think and why you behave the way you behave. I also hope that some readers will be inspired to focus their careers on building intelligent machines based on the principles outlined in these pages.

I often refer to this theory and my approach to studying intelligence as "real intelligence" to distinguish it from "artificial intelligence." AI scientists tried to program computers to act like humans without first answering what intelligence is and what it means to understand. They left out the most important part of building intelligent machines, the intelligence! "Real intelligence" makes the point that before we attempt to build intelligent machines, we have to first understand how the brain thinks, and there is nothing artificial about that. Only then can we ask how we can build intelligent machines.

The book starts with some background on why previous attempts at understanding intelligence and building intelligent machines have failed. I then introduce and develop the core idea of the theory, what I call the memory-prediction framework. In chapter 6 I detail how the physical brain implements the memory-prediction model—in other words, how the brain actually works. I then discuss social and other implications of the theory, which for many readers might be the most thought-provoking section. The book ends with a discussion of intelligent machines—how we can build them and what the future will be like. I hope you find it fascinating. Here are some of the questions we will cover along the way:

> ***Can computers be intelligent?***
> For decades, scientists in the field of artificial intelligence have claimed that computers will be intelligent when they are powerful enough. I don't think so, and I will explain why. Brains and computers do fundamentally different things.
>
> ***Weren't neural networks supposed to lead to intelligent machines?***
> Of course the brain is made from a network of neurons, but without first understanding what the brain does, simple neural networks will be no more successful at creating intelligent machines than computer programs have been.
>
> ***Why has it been so hard to figure out how the brain works?***
> Most scientists say that because the brain is so complicated, it will take a very long time for us to understand it. I disagree. Complexity is a symptom of confusion, not a cause. Instead, I argue we have a few intuitive but incorrect

assumptions that mislead us. The biggest mistake is the belief that intelligence is defined by intelligent behavior.

What is intelligence if it isn't defined by behavior?
The brain uses vast amounts of memory to create a model of the world. Everything you know and have learned is stored in this model. The brain uses this memory-based model to make continuous predictions of future events. It is the ability to make predictions about the future that is the crux of intelligence. I will describe the brain's predictive ability in depth; it is the core idea in the book.

How does the brain work?
The seat of intelligence is the neocortex. Even though it has a great number of abilities and powerful flexibility, the neocortex is surprisingly regular in its structural details. The different parts of the neocortex, whether they are responsible for vision, hearing, touch, or language, all work on the same principles. The key to understanding the neocortex is understanding these common principles and, in particular, its hierarchical structure. We will examine the neocortex in sufficient detail to show how its structure captures the structure of the world. This discussion will be the most technical part of the book, but interested nonscientist readers should be able to understand it.

What are the implications of this theory?
This theory of the brain can help explain many things, such as how we are creative, why we feel conscious, why we exhibit prejudice, how we learn, and why "old dogs" have trouble learning "new tricks." I will discuss a number of these topics. Overall, this theory gives us insight into who we are and why we do what we do.

Can we build intelligent machines and what will they do?
Yes. We can and we will. Over the next few decades, I see the capabilities of such machines evolving rapidly and in interesting directions. Some people fear that intelligent machines could be dangerous to humanity, but I argue strongly against this idea. We are not going to be overrun by robots. It will be far easier to build machines that outstrip our abilities in high-level thought such as physics and mathematics than to build anything like the walking, talking robots we see in popular fiction. I will explore the incredible directions in which this technology is likely to go.

My goal is to explain this new theory of intelligence and how the brain works in a way that anybody will be able to understand. A good theory should be easy to comprehend, not obscured in jargon or convoluted argument. I'll start with a basic framework and then add details as we go. Some will be reasoning just on logical grounds; some will involve particular aspects of brain circuitry. Some of the details of what I propose are certain to be wrong, which is always the case in any area of science. A fully mature theory will take years to develop, but that doesn't diminish the power of the core idea.

When I first became interested in brains many years ago, I went to my local library to look for a good book that would explain how brains worked. As a teenager I had become accustomed to being able to find well-written books that explained almost any topic of interest. There were books on relativity theory, black holes, magic, and mathematics—whatever I was fascinated with at the moment. Yet my search for a satisfying brain book turned up empty. I came to realize that no one had any idea how the brain actually worked. There weren't even any bad or unproven theories; there simply were none. This was unusual. For example, at that time no one knew how the dinosaurs had died, but there were plenty of theories, all of which you could read about. There was nothing like this for brains. At first I had trouble believing it. It bothered me that we didn't know how this critical organ worked. While studying what we did know about brains, I came to believe that there must be a straightforward explanation. The brain wasn't magic, and it didn't seem to me that the answers would even be that complex. The mathematician Paul Erdös believed that the simplest mathematical proofs already exist in some ethereal book and a mathematician's job was to find them, to "read the book." In the same way, I felt that the explanation of intelligence was "out there." I could taste it. I wanted to read the book.

For the past twenty-five years, I have had a vision of that small, straightforward book on the brain. It was like a carrot keeping me motivated during those years. This vision has shaped the book you are holding in your hands right now. I have never liked complexity, in either science or technology. You can see that reflected in the products I have designed, which are often noted for their ease of use. The most powerful things are simple. Thus this book proposes a simple and straightforward theory of intelligence. I hope you enjoy it.

1

ARTIFICIAL INTELLIGENCE

When I graduated from Cornell in June 1979 with a degree in electrical engineering, I didn't have any major plans for my life. I started work as an engineer at the new Intel campus in Portland, Oregon. The microcomputer industry was just starting, and Intel was at the heart of it. My job was to analyze and fix problems found by other engineers working in the field with our main product, single board computers. (Putting an entire computer on a single circuit board had only recently been made possible by Intel's invention of the microprocessor.) I published a newsletter, got to do some traveling, and had a chance to meet customers. I was young and having a good time, although I missed my college sweetheart who had taken a job in Cincinnati.

A few months later, I encountered something that was to change my life's direction. That something was the newly published September issue of *Scientific American,* which was dedicated entirely to the brain. It rekindled my childhood interest in brains. It was fascinating. From it I learned about the organization, development, and chemistry of the brain, neural mechanisms of vision, movement, and other specializations, and the biological basis for disorders of the mind. It was one of the best *Scientific American* issues of all time. Several neuroscientists I've spoken to have told me it played a significant role in their career choice, just as it did for me.

The final article, "Thinking About the Brain," was written by Francis Crick, the codiscoverer of the structure of DNA who had by then turned his talents to studying the brain. Crick argued that in spite of a steady accumulation of detailed knowledge about the brain, how the brain worked was still a profound mystery. Scientists usually don't write about what they don't know, but Crick didn't care. He was like the boy pointing to the emperor with no clothes. According to Crick, neuroscience was a lot of data without a theory. His exact words were, "what is conspicuously lacking is a broad framework of ideas." To me this was the British gentleman's way of saying, "We don't have a clue how this thing works." It was true then, and it's still true today.

Crick's words were to me a rallying call. My lifelong desire to understand brains and build intelligent machines was brought to life. Although I was barely out of college, I decided to change careers. I was going to study brains, not only to understand how they worked, but to use that knowledge as a foundation for new technologies, to build intelligent machines. It would take some time to put this plan into action.

In the spring of 1980 I transferred to Intel's Boston office to be reunited with my future wife, who was starting graduate school. I took a position teaching customers and employees how to design microprocessor-based systems. But I had my sights on a different goal: I was trying to figure out how to work on brain theory. The engineer in me realized that once we understood how brains worked, we could build them, and the natural way to build artificial brains was in silicon. I worked for the company that invented the silicon memory chip and the microprocessor; therefore, perhaps I could interest Intel in letting me spend part of my time thinking about intelligence and how we could design brainlike memory chips. I wrote a letter to Intel's chairman, Gordon Moore. The letter can be distilled to the following:

> Dear Dr. Moore,
> I propose that we start a research group devoted to understanding how the brain works. It can start with one person—me—and go from there. I am confident we can figure this out. It will be a big business one day.
>
> —Jeff Hawkins

Moore put me in touch with Intel's chief scientist, Ted Hoff. I flew to California to meet him and lay out my proposal for studying the brain. Hoff was famous for two things. The first, which I was aware of, was for his work in designing the first microprocessor. The second, which I was not aware of at the time, was for his work in early neural network theory. Hoff had experience with artificial neurons and some of the things you could do with them. I wasn't prepared for this. After listening to my proposal, he said he didn't believe it would be possible to figure out how the brain works in the foreseeable future, and so it didn't make sense for Intel to support me. Hoff was correct, because it is now twenty-five years later and we are just starting to make significant progress in understanding brains. Timing is everything in business. Still, at the time I was pretty disappointed.

I tend to seek the path of least friction to achieve my goals. Working on brains at Intel would have been the simplest transition. With that option eliminated I looked for the next best thing. I decided to apply to graduate school at the Massachusetts Institute of Technology, which was famous for its research on artificial intelligence and was conveniently located down the road. It seemed a great match. I had extensive training in computer science—"check." I had a desire to build intelligent machines, "check." I wanted to first study brains to see how they worked . . . "uh, that's a problem." This last goal, wanting to understand how brains worked, was a nonstarter in the eyes of the scientists at the MIT artificial intelligence lab.

It was like running into a brick wall. MIT was the mothership of artificial intelligence. At the time I applied to MIT, it was home to dozens of bright people who were enthralled with the idea of programming computers to produce intelligent behavior. To these scientists, vision, language, robotics, and mathematics were just programming problems. Computers could do anything a brain could do, and more, so why constrain your thinking by the biological messiness of nature's computer? Studying brains would limit your thinking. They believed it was better to study the ultimate limits of computation as best expressed in digital computers. Their holy grail was to write computer programs that would first match and then surpass human abilities. They took an ends-justify-the-means approach; they were not interested in how real brains worked. Some took pride in ignoring neurobiology.

This struck me as precisely the wrong way to tackle the problem. Intuitively I felt that the artificial intelligence approach would not only fail to create programs that do what humans can do, it would not teach us what intelligence is. Computers and brains are built on completely different principles. One is programmed, one is self-learning. One has to be perfect to work at all, one is naturally flexible and tolerant of failures. One has a central processor, one has no centralized control. The list of differences goes on and on. The biggest reason I thought computers would not be intelligent is that I understood how computers worked, down to the level of the transistor physics, and this knowledge gave me a strong intuitive sense that brains and computers were fundamentally different. I couldn't prove it, but I knew it as much as one can intuitively know anything.

Ultimately, I reasoned, AI might lead to useful products, but it wasn't going to build truly intelligent machines.

In contrast, I wanted to understand real intelligence and perception, to study brain physiology and anatomy, to meet Francis Crick's challenge and come up with a broad framework for how the brain worked. I set my sights in particular on the neocortex—the most recently developed part of the mammalian brain and the seat of intelligence. After understanding how the neocortex worked, then we could go about building intelligent machines, but not before.

Unfortunately, the professors and students I met at MIT did not share my interests. They didn't believe that you needed to study real brains to understand intelligence and build intelligent machines. They told me so. In 1981 the university rejected my application.

Many people today believe that AI is alive and well and just waiting for enough computing power to deliver on its many promises. When computers have sufficient memory and processing power, the thinking goes, AI programmers will be able to make intelligent machines. I disagree. AI suffers from a fundamental flaw in that it fails to adequately address what intelligence is or what it means to understand something. A brief look at the history of AI and the tenets on which it was built will explain how the field has gone off course.

The AI approach was born with the digital computer. A key figure in the early AI movement was the English mathematician Alan Turing, who was one of the inventors of the idea of the general-purpose computer. His masterstroke was to formally demonstrate the concept of universal computation: that is, all computers are fundamentally equivalent regardless of the details of how they are built. As part of his proof, he conceived an imaginary machine with three essential parts: a processing box, a paper tape, and a device that reads and writes marks on the tape as it moves back and forth. The tape was for storing information—like the famous 1's and 0's of computer code (this was before the invention of memory chips or the disk drive, so Turing imagined paper tape for storage). The box, which today we call a central processing unit (CPU), follows a set of fixed rules for reading and editing the information on the tape. Turing proved, mathematically, that if you choose the right set of rules for the CPU and give it an indefinitely long tape to work with, it can perform any definable set of operations in the universe. It would be one of many equivalent machines now called Universal Turing Machines. Whether the problem is to compute square roots, calculate ballistic trajectories, play games, edit pictures, or reconcile bank transactions, it is all 1's and 0's underneath, and any Turing Machine can be programmed to handle it. Information processing is information processing is information processing. All digital computers are logically equivalent.

Turing's conclusion was indisputably true and phenomenally fruitful. The computer revolution and all its products are built on it. Then Turing turned to the question of how to build an intelligent machine. He felt computers could be intelligent, but he didn't want to get into arguments about whether this was possible or not. Nor did he think he could define *intelligence* formally, so he didn't even try. Instead, he proposed an existence proof for intelligence, the famous Turing Test: if a computer can fool a human interrogator into thinking that it too is a person, then by definition the computer must be intelligent. And so, with the Turing Test as his measuring stick and the Turing Machine as his medium, Turing helped launch the field of AI. Its central dogma: the brain is just another kind of computer. It doesn't matter how you design an artificially intelligent system, it just has to produce humanlike behavior.

The AI proponents saw parallels between computation and

thinking. They said, "Look, the most impressive feats of human intelligence clearly involve the manipulation of abstract symbols—and that's what computers do too. What do we do when we speak or listen? We manipulate mental symbols called words, using well-defined rules of grammar. What do we do when we play chess? We use mental symbols that represent the properties and locations of the various pieces. What do we do when we see? We use mental symbols to represent objects, their positions, their names, and other properties. Sure, people do all this with brains and not with the kinds of computers we build, but Turing has shown that it doesn't matter how you implement or manipulate the symbols. You can do it with an assembly of cogs and gears, with a system of electronic switches, or with the brain's network of neurons—whatever, as long as your medium can realize the functional equivalent of a Universal Turing Machine."

This assumption was bolstered by an influential scientific paper published in 1943 by the neurophysiologist Warren McCulloch and the mathematician Walter Pitts. They described how neurons could perform digital functions—that is, how nerve cells could conceivably replicate the formal logic at the heart of computers. The idea was that neurons could act as what engineers call logic gates. Logic gates implement simple logical operations such as AND, NOT, and OR. Computer chips are composed of millions of logic gates all wired together into precise, complicated circuits. A CPU is just a collection of logic gates.

McCulloch and Pitts pointed out that neurons could also be connected together in precise ways to perform logic functions. Since neurons gather input from each other and process those inputs to decide whether to fire off an output, it was conceivable that neurons might be living logic gates. Thus, they inferred, the brain could conceivably be built out of AND-gates, OR-gates, and other logic elements all built with neurons, in direct analogy with the wiring of digital electronic circuits. It isn't clear whether McCulloch and Pitts actually believed the brain worked this way; they only said it was possible. And, logically speaking, this view of neurons is possible. Neurons can, in theory, implement digital functions. However, no one bothered to ask if that was how neurons actually were wired in the brain. They took it as proof, irrespective of the lack of biological evidence, that brains were just another kind of computer.

It's also worth noting that AI philosophy was buttressed by the dominant trend in psychology during the first half of the twentieth century, called behaviorism. The behaviorists believed that it was not possible to know what goes on inside the brain, which they called an impenetrable black box. But one could observe and measure an animal's environment and its behaviors—what it senses and what it does, its inputs and its outputs. They conceded that the brain contained reflex mechanisms that could be used to condition an animal into adopting new behaviors through reward and punishments. But other than this, one did not need to study the brain, especially messy subjective feelings such as hunger, fear, or what it means to understand something. Needless to say, this research philosophy eventually withered away throughout the second half of the twentieth century, but AI would stick around a lot longer.

As World War II ended and electronic digital computers became available for broader applications, the pioneers of AI rolled up their sleeves and began programming. Language translation? Easy! It's a kind of code breaking. We just need to map each symbol in System A onto its counterpart in System B. Vision? That looks easy too. We already know geometric theorems that deal with rotation, scale, and displacement, and we can easily encode them as computer algorithms—so we're halfway there. AI pundits made grand claims about how quickly computer intelligence would first match and then surpass human intelligence.

Ironically, the computer program that came closest to passing the Turing Test, a program called Eliza, mimicked a psychoanalyst, rephrasing your questions back at you. For example, if a person typed in, "My boyfriend and I don't talk anymore," Eliza might say, "Tell me more about your boyfriend" or "Why do you think your boyfriend and you don't talk anymore?" Designed as a joke, the program actually fooled some people, even though it was dumb and trivial. More serious efforts included programs such as Blocks World, a simulated room containing blocks of different colors and shapes. You could pose questions to Blocks World such as "Is there a green pyramid on top of the big red cube?" or "Move the blue cube on top of the little red cube." The program would answer your question or try to do what you asked. It was all simulated—and it worked. But it was limited to its own highly artificial world of blocks. Programmers couldn't generalize it to do anything useful.

The public, meanwhile, was impressed by a continuous stream of seeming successes and news stories about AI technology. One program that generated initial excitement was able to solve mathematical theorems. Ever since Plato, multistep deductive inference has been seen as the pinnacle of human intelligence, so at first it seemed that AI had hit the jackpot. But, like Blocks World, it turned out the program was limited. It could only find very simple theorems, which were already known. Then there was a large stir about "expert systems," databases of facts that could answer questions posed by human users. For example, a medical expert system might be able to diagnose a patient's disease if given a list of symptoms. But again, they turned out to be of limited use and didn't exhibit anything close to generalized intelligence. Computers could play checkers at expert skill levels and eventually IBM's Deep Blue famously beat Gary Kasparov, the world chess champion, at his own game. But these successes were hollow. Deep Blue didn't win by being smarter than a human; it won by being millions of times faster than a human. Deep Blue had no intuition. An expert human player looks at a board position and immediately sees what areas of play are most likely to be fruitful or dangerous, whereas a computer has no innate sense of what is important and must explore many more options. Deep Blue also had no sense of the history of the game, and didn't know anything about its opponent. It played chess yet didn't understand chess, in the same way that a calculator performs arithmetic but doesn't understand mathematics.

In all cases, the successful AI programs were only good at the one particular thing for which they were specifically designed. They didn't generalize or show flexibility, and even their creators admitted they didn't think like humans. Some AI problems, which at first were thought to be easy, yielded no progress. Even today, no computer can understand language as well as a three-year-old or see as well as a mouse.

After many years of effort, unfulfilled promises, and no unqualified successes, AI started to lose its luster. Scientists in the field moved on to other areas of research. AI start-up companies failed. And funding became scarcer. Programming computers to do even the most basic tasks of perception, language, and behavior began to seem impossible. Today, not much has changed. As I said earlier, there are still people who believe that AI's problems can be solved with faster computers, but most scientists think the entire endeavor was flawed.

We shouldn't blame the AI pioneers for their failures. Alan Turing was brilliant. They all could tell that the Turing Machine would change the world—and it did, but not through AI.

My skepticism of AI's assertions was honed around the same time that I applied to MIT. John Searle, an influential philosophy professor at the University of California at Berkeley, was at that time saying that computers were not, and could not be, intelligent. To prove it, in 1980 he came up with a thought experiment called the Chinese Room. It goes like this:

Suppose you have a room with a slot in one wall, and inside is an English-speaking person sitting at a desk. He has a big book of instructions and all the pencils and scratch paper he could ever need. Flipping through the book, he sees that the instructions, written in English, dictate ways to manipulate, sort, and compare Chinese characters. Mind you, the directions say nothing about the meanings of the Chinese characters; they only deal with how the characters are to be copied, erased, reordered, transcribed, and so forth.

Someone outside the room slips a piece of paper through the slot. On it is written a story and questions about the story, all in Chinese. The man inside doesn't speak or read a word of Chinese, but he picks up the paper and goes to work with the rulebook. He toils and toils, rotely following the instructions in the book. At times the instructions tell him to write characters on scrap paper, and at other times to move and erase characters. Applying rule after rule, writing and erasing characters, the man works until the book's instructions tell him he is done. When he is finished at last he has written a new page of characters, which unbeknownst to him are the answers to the questions. The book tells him to pass his paper back through the slot. He does it, and wonders what this whole tedious exercise has been about.

Outside, a Chinese speaker reads the page. The answers are all correct, she notes—even insightful. If she is asked whether those answers came from an intelligent mind that had understood the story, she will definitely say yes. But can she be right? Who understood the story? It wasn't the fellow inside, certainly; he is ignorant of Chinese and has no idea what the story was about. It wasn't the book, which is just, well, a book, sitting inertly on the writing desk amid piles of paper. So where did the understanding occur? Searle's answer is that no understanding did occur; it was just a bunch of mindless page flipping and pencil scratching. And now the bait-and-switch: the Chinese Room is exactly analogous to a digital computer. The person is the CPU, mindlessly executing instructions, the book is the software program feeding instructions to the CPU, and the scratch paper is the memory. Thus, no matter how cleverly a computer is designed to simulate intelligence by producing the same behavior as a human, it has no understanding and it is not intelligent. (Searle made it clear he didn't know what intelligence is; he was only saying that whatever it is, computers don't have it.)

This argument created a huge row among philosophers and AI pundits. It spawned hundreds of articles, plus more than a little vitriol and bad blood. AI defenders came up with dozens of counterarguments to Searle, such as claiming that although none of the room's component parts understood Chinese, the entire room as a whole did, or that the person in the room really did understand Chinese, but just didn't know it. As for me, I think Searle had it right. When I thought through the Chinese Room argument and when I thought about how computers worked, I didn't see understanding happening anywhere. I was convinced we needed to understand what "understanding" is, a way to define it that would make it clear when a system was intelligent and when it wasn't, when it understands Chinese and when it doesn't. Its behavior doesn't tell us this.

A human doesn't need to "do" anything to understand a story. I can read a story quietly, and although I have no overt behavior my understanding and comprehension are clear, at least to me. You, on the other hand, cannot tell from my quiet behavior whether I understand the story or not, or even if I know the language the story is written in. You might later ask me questions to see if I did, but my understanding occurred when I read the story, not just when I answer your questions. A thesis of this book is that understanding cannot be measured by external behavior; as we'll see in the coming chapters, it is instead an internal metric of how the brain remembers things and uses its memories to make predictions. The Chinese Room, Deep Blue, and most computer programs don't have anything akin to this. They don't understand what they are doing. The only way we can judge whether a computer is intelligent is by its output, or behavior.

The ultimate defensive argument of AI is that computers could, in theory, simulate the entire brain. A computer could model all the neurons and their connections, and if it did there would be nothing to distinguish the "intelligence" of the brain from the "intelligence" of the computer simulation. Although this may be impossible in practice, I agree with it. But AI researchers don't simulate brains, and their programs are not intelligent. You can't simulate a brain without first understanding what it does.

After my rejection by both Intel and MIT, I didn't know what to do. When you don't know how to proceed, often the best strategy is to make no changes until your options become clear. So I just kept working in the computer field. I was content to stay in Boston, but in 1982 my wife wanted to move to California, so we did (it was, again, the path of least friction). I landed a job in Silicon Valley, at a start-up called Grid Systems. Grid invented the laptop computer, a beautiful machine that became the first computer in the collection at the Museum of Modern Art in New York. Working first in marketing and then in engineering, I eventually created a high-level programming language called GridTask. It and I became more and more important to Grid's success; my career was going well.

Still, I could not get my curiosity about the brain and intelligent machines out of my head. I was consumed with a desire to study brains. So I took a correspondence course in human physiology and studied on my own (no one ever got rejected by a correspondence school!). After learning a fair amount of biology, I decided to apply for admission to a biology graduate program and study intelligence from within the biological sciences. If the computer science world didn't want a brain theorist, then maybe the biology world would welcome a computer scientist. There was no such thing as theoretical biology back then, and especially not theoretical neuroscience, so biophysics seemed like the best field for my interests. I studied hard, took the required entrance exams, prepared a résumé, solicited letters of recommendation, and voilà, I was accepted as a full-time graduate student in the biophysics program at the University of California at Berkeley.

I was thrilled. Finally I could start in earnest on brain theory, or so I thought. I quit my job at Grid with no intention of working in the computer industry again. Of course this meant indefinitely giving up my salary. My wife was thinking "time to buy a house and start a family" and I was happily becoming a nonprovider. This was definitely not a low-friction path. But it was the best option I had, and she supported my decision.

John Ellenby, the founder of Grid, pulled me into his office just before I left and said, "I know you don't expect to ever come back to Grid or the computer industry, but you never know what will happen. Instead of stopping completely, why don't you take a leave of absence? That way if, in a year or two, you do come back, you can pick up your salary, position, and stock options where you left off." It was a nice gesture. I accepted it, but I felt I was leaving the computer business for good.

2

NEURAL NETWORKS

When I started at UC Berkeley in January 1986, the first thing I did was compile a history of theories of intelligence and brain function. I read hundreds of papers by anatomists, physiologists, philosophers, linguists, computer scientists, and psychologists. Numerous people from many fields had written extensively about thinking and intelligence. Each field had its own set of journals and each used its own terminology. I found their descriptions inconsistent and incomplete. Linguists talked of intelligence in terms such as "syntax" and "semantics." To them, the brain and intelligence was all about language. Vision scientists referred to 2D, 2½D, and 3D sketches. To them, the brain and intelligence was all about visual pattern recognition. Computer scientists talked of schemas and frames, new terms they made up to represent knowledge. None of these people talked about the structure of the brain and how it would implement any of their theories. On the other hand, anatomists and neurophysiologists wrote extensively about the structure of the brain and how neurons behave, but they mostly avoided any attempt at large-scale theory. It was difficult and frustrating trying to make sense of these various approaches and the mountain of experimental data that accompanied them.

Around this time, a new and promising approach to thinking about intelligent machines burst onto the scene. Neural networks had been around since the late 1960s in one form or another, but neural networks and the AI movement were competitors, for both the dollars and the mind share of the agencies that fund research. AI, the 800-pound gorilla in those days, actively squelched neural network research. Neural network researchers were essentially blacklisted from getting funding for several years. A few people continued to think about them though, and in the mid-1980s their day in the sun had finally arrived. It is hard to know exactly why there was a sudden interest in neural networks, but undoubtedly one contributing factor was the continuing failure of artificial intelligence. People were casting about for alternatives to AI and found one in artificial neural networks.

Neural networks were a genuine improvement over the AI approach because their architecture is based, though very loosely, on real nervous systems. Instead of programming computers, neural network researchers, also known as *connectionists,* were interested in learning what kinds of behaviors could be exhibited by hooking a bunch of neurons together. Brains are made of neurons; therefore, the brain is a neural network. That is a fact. The hope of connectionists was that the elusive properties of intelligence would become clear by studying how neurons interact, and that some of the problems that were unsolvable with AI could be solved by replicating the correct connections between populations of neurons. A neural network is unlike a computer in that it has no CPU and doesn't store information in a centralized memory. The network's knowledge and memories are distributed throughout its connectivity—just like real brains.

On the surface, neural networks seemed to be a great fit with my own interests. But I quickly became disillusioned with the field. By this time I had formed an opinion that three things were essential to understanding the brain. My first criterion was the inclusion of time in brain function. Real brains process rapidly changing streams of information. There is nothing static about the flow of information into and out of the brain.

The second criterion was the importance of feedback. Neuroanatomists have known for a long time that the brain is saturated with feedback connections. For example, in the circuit between the neocortex and a lower structure called the thalamus, connections going backward (toward the input) exceed the connections going forward by almost a factor of ten! That is, for every fiber feeding information forward into the neocortex, there are ten fibers feeding information back toward the senses. Feedback dominates most connections throughout the neocortex as well. No one understood the precise role of this feedback, but it was clear from published research that it existed everywhere. I figured it must be important.

The third criterion was that any theory or model of the brain should account for the physical architecture of the brain. The neocortex is not a simple structure. As we will see later, it is organized as a repeating hierarchy. Any neural network that didn't acknowledge this structure was certainly not going to work like a brain.

But as the neural network phenomenon exploded on the scene, it mostly settled on a class of ultrasimple models that didn't meet any of these criteria. Most neural networks consisted of a small number of neurons connected in three rows. A pattern (the input) is presented to the first row. These input neurons are connected to the next row of neurons, the so-called hidden units. The hidden units then connect to the final row of neurons, the output units. The connections between neurons have variable strengths, meaning the activity in one neuron might increase the activity in another and decrease the activity in a third neuron depending on the connection strengths. By changing these strengths, the network learns to map input patterns to output patterns.

These simple neural networks only processed static patterns, did not use feedback, and didn't look anything like brains. The most common type of neural network, called a "back propagation" network, learned by broadcasting an error from the output units back toward the input units. You might think this is a form of feedback, but it isn't really. The backward propagation of errors only occurred during the learning phase. When the neural network was working normally, after being trained, the information flowed only one way. There was no feedback from outputs to inputs. And the models had no sense of time. A static input pattern got converted into a static output pattern. Then another input pattern was presented. There was no history or record in the network of what happened even a short time earlier. And finally the architecture of these neural networks was trivial compared to the complicated and hierarchical structure of the brain.

I thought the field would quickly move on to more realistic networks, but it didn't. Because these simple neural networks were able to do interesting things, research seemed to stop right there, for years. They had found a new and interesting tool, and overnight thousands of scientists, engineers, and students were getting grants, earning PhDs, and writing books about neural networks. Companies were formed to use neural networks to predict the stock market, process loan applications, verify signatures, and perform hundreds of other pattern classification applications. Although the intent of the founders of the field might have been more general, the field became dominated by people who weren't interested in understanding how the brain works, or understanding what intelligence is.

The popular press didn't understand this distinction well.

Newspapers, magazines, and TV science programs presented neural networks as being "brainlike" or working on the "same principles as the brain." Unlike AI, where everything had to be programmed, neural nets learned by example, which seemed, well, somehow more intelligent. One prominent demonstration was NetTalk. This neural network learned to map sequences of letters onto spoken sounds. As the network was trained on printed text, it started sounding like a computer voice reading the words. It was easy to imagine that, with a little more time, neural networks would be conversing with humans. NetTalk was incorrectly heralded on national news as a machine learning to read. NetTalk was a great exhibition, but what it was actually doing bordered on the trivial. It didn't read, it didn't understand, and was of little practical value. It just matched letter combinations to predefined sound patterns.

Let me give you an analogy to show how far neural networks were from real brains. Imagine that instead of trying to figure out how a brain worked we were trying to figure out how a digital computer worked. After years of study, we discover that everything in the computer is made of transistors. There are hundreds of millions of transistors in a computer and they are connected together in precise and complex ways. But we don't understand how the computer works or why the transistors are connected the way they are. So one day we decide to connect just a few transistors together to see what happens. Lo and behold we find that as few as three transistors, when connected together in a certain way, become an amplifier. A small signal put into one end is magnified on the other end. (Amplifiers in radios and televisions are made using transistors in this fashion.) This is an important discovery, and overnight an industry springs up making transistor radios, televisions, and other electronic appliances using transistor amplifiers. This is all well and good, but it doesn't tell us anything about how the computer works. Even though an amplifier and a computer are both made of transistors, they have almost nothing else in common. In the same way, a real brain and a three-row neural network are built with neurons, but have almost nothing else in common.

During the summer of 1987, I had an experience that threw more cold water on my already low enthusiasm for neural nets. I went to a neural network conference where I saw a presentation by a company called Nestor. Nestor was trying to sell a neural network application for recognizing handwriting on a tablet. It was offering to license the program for one million dollars. That got my attention. Although Nestor was promoting the sophistication of its neural network algorithm and touting it as yet another major breakthrough, I felt the problem of handwriting recognition could be solved in a simpler, more traditional way. I went home that night, thought about the problem, and in two days had designed a handwriting recognizer that was fast, small, and flexible. My solution didn't use a neural network and it didn't work at all like a brain. Although that conference sparked my interest in designing computers with a stylus interface (eventually leading to the PalmPilot ten years later), it also convinced me that neural networks were not much of an improvement over traditional methods. The handwriting recognizer I created ultimately became the basis for the text entry system, called Graffiti, used in the first series of Palm products. I think Nestor went out of business.

So much for simple neural networks. Most of their capabilities were easily handled by other methods and eventually the media hoopla subsided. At least neural network researchers did not claim their models were intelligent. After all, they were extremely simple networks and did less than AI programs. I don't want to leave you with the impression that all neural networks are of the simple three-layer variety. Some researchers have continued to study neural networks of different designs. Today the term *neural network* is used to describe a diverse set of models, some of which are more biologically accurate and some of which are not. But almost none of them attempt to capture the overall function or architecture of the neocortex.

In my opinion, the most fundamental problem with most neural networks is a trait they share with AI programs. Both are fatally burdened by their focus on behavior. Whether they are calling these behaviors "answers," "patterns," or "outputs," both AI and neural networks assume intelligence lies in the behavior that a program or a neural network produces after processing a given input. The most important attribute of a computer program or a neural network is whether it gives the correct or desired output. As inspired by Alan Turing, intelligence equals behavior.

But intelligence is not just a matter of acting or behaving intelligently. Behavior is a manifestation of intelligence, but not the central characteristic or primary definition of being intelligent. A moment's reflection proves this: You can be intelligent just lying in the dark, thinking and understanding. Ignoring what goes on *in* your head and focusing instead on behavior has been a large impediment to understanding intelligence and building intelligent machines.

Before we explore a new definition of intelligence, I want to tell you about one other connectionist approach that came much closer to describing how real brains work. Trouble is, few people seem to have realized the importance of this research.

While neural nets grabbed the limelight, a small splinter group of neural network theorists built networks that didn't focus on behavior. Called auto-associative memories, they were also built out of simple "neurons" that connected to each other and fired when they reached a certain threshold. But they were interconnected differently, using lots of feedback. Instead of only passing information forward, as in a back propagation network, auto-associative memories fed the output of each neuron back into the input—sort of like calling yourself on the phone. This feedback loop led to some interesting features. When a pattern of activity was imposed on the artificial neurons, they formed a memory of this pattern. The auto-associative network associated patterns with themselves, hence the term *auto-associative memory*.

The result of this wiring may at first seem ridiculous. To retrieve a pattern stored in such a memory, you must provide the pattern you want to retrieve. It would be like going to the grocer and asking to buy a bunch of bananas. When the grocer asks you how you will pay, you offer to pay with bananas. What good is that? you might ask. But an auto-associative memory has a few important properties that are found in real brains.

The most important property is that you don't have to have the entire pattern you want to retrieve in order to retrieve it. You might have only part of the pattern, or you might have a somewhat messed-up pattern. The auto-associative memory can retrieve the correct pattern, as it was originally stored, even though you start with a messy version of it. It would be like going to the grocer with half eaten brown bananas and getting whole green bananas in return. Or going to the bank with a ripped and unreadable bill and the banker says, "I think this is a messed-up $100 bill. Give me that one, and I will give you this new, crisp $100 bill."

Second, unlike most other neural networks, an auto-associative memory can be designed to store sequences of patterns, or temporal patterns. This feature is accomplished by adding a time delay to the feedback. With this delay, you can present an auto-associative memory with a sequence of patterns, similar to a melody, and it can remember the sequence. I might feed in the first few notes of "Twinkle Twinkle Little Star" and the memory returns the whole song. When presented with part of the sequence, the memory can recall the rest. As we will see later, this is how people learn practically everything, as a sequence of patterns. And I propose the brain uses circuits similar to an auto-associative memory to do so.

Auto-associative memories hinted at the potential importance of feedback and time-changing inputs. But the vast majority of AI, neural network, and cognitive scientists ignored time and feedback.

Neuroscientists as a whole have not done much better. They too know about feedback—they are the people who discovered it—but most have no theory (beyond vague talk of "phases" and "modulation") to account for why the brain needs to have so much of it. And time has little or no central role in most of their ideas on overall brain function. They tend to chart the brain in terms of where things happen, not when or how neural firing patterns interact over time. Part of this bias comes from the limits of our current experimental techniques. One of the favorite technologies of the 1990s, aka the Decade of the Brain, was functional imaging. Functional imaging machines can take pictures of brain activity in humans. However, they cannot see rapid changes. So scientists ask subjects to concentrate on a single task over and over again as if they were being asked to stand still for an optical photograph, except this is a mental photograph. The result is we have lots of data on *where* in the brain certain tasks occur, but little data on how realistic, time-varying inputs flow through the brain. Functional imaging lends insight into *where* things are happening at a given moment but cannot easily capture how brain activity changes over time. Scientists would like to collect this data, but there are few good techniques for doing so. Thus many mainstream cognitive neuroscientists continue to buy into the input-output fallacy. You present a fixed input and see what output you get. Wiring diagrams of the cortex tend to show flowcharts that start in the primary sensory areas where sights, sounds, and touch come in, flow up through higher analytical, planning, and motor areas, and then feed instructions down to the muscles. You sense, then you act.

I don't want to imply that everyone has ignored time and feedback. This is such a big field that virtually every idea has its adherents. In recent years, belief in the importance of feedback, time, and prediction has been on the rise. But the thunder of AI and classical neural networks kept other approaches subdued and underappreciated for many years.

It's not difficult to understand why people—laymen and experts alike—have thought that behavior defines intelligence. For at least a couple of centuries people have likened the brain's abilities to clockworks, then pumps and pipes, then steam engines and, later, to computers. Decades of science fiction have been awash in AI ideas, from Isaac Asimov's laws of robotics to *Star Wars*' C3PO. The idea of intelligent machines *doing* things is engrained in our imagination. All machines, whether made by humans or imagined by humans, are designed to do something. We don't have machines that think, we have machines that do. Even as we observe our fellow humans, we focus on their behavior and not on their hidden thoughts. Therefore, it seems intuitively obvious that intelligent behavior should be the metric of an intelligent system.

However, looking across the history of science, we see our intuition is often the biggest obstacle to discovering the truth. Scientific frameworks are often difficult to discover, not because they are complex, but because intuitive but incorrect assumptions keep us from seeing the correct answer. Astronomers before Copernicus (1473–1543) wrongly assumed that the earth was stationary at the center of the universe because it *feels* stationary and *appears* to be at the center of the universe. It was intuitively obvious that the stars were all part of a giant spinning sphere, with us at its center. To suggest the Earth was spinning like a top, the surface moving at nearly a thousand miles an hour, and that the entire Earth was hurtling through space—not to mention that stars are trillions of miles away—would have marked you as a lunatic. But that turned out to be the correct framework. Simple to understand, but intuitively incorrect.

Before Darwin (1809–1882), it seemed obvious that species are fixed in their forms. Crocodiles don't mate with hummingbirds; they are distinct and irreconcilable. The idea that species evolve went against not only religious teachings but also common sense. Evolution implies that you have a common ancestor with every living thing on this planet, including worms and the flowering plant in your kitchen. We now know this to be true, but intuition says otherwise.

I mention these famous examples because I believe that the quest for intelligent machines has also been burdened by an intuitive assumption that's hampering our progress. When you ask yourself, *What does an intelligent system do?*, it is intuitively obvious to think in terms of behavior. We demonstrate human intelligence through our speech, writing, and actions, right? Yes, but only to a point. Intelligence is something that is happening in your head. Behavior is an optional ingredient. This is not intuitively obvious, but it's not hard to understand either.

In the spring of 1986, as I sat at my desk day after day reading scientific articles, building my history of intelligence, and watching the evolving worlds of AI and neural networks, I found myself drowning in details. There was an unending supply of things to study and read about, but I was not gaining any clear understanding of how the whole brain actually worked or even what it did. This was because the field of neuroscience itself was awash in details. It still is. Thousands of research reports are published every year, but they tend to add to the heap rather than organize it. There's still no overall theory, no framework, explaining what your brain does and how it does it.

I started imagining what the solution to this problem would be like. Is it going to be extremely complex because the brain is so complex? Would it take one hundred pages of dense mathematics to describe how the brain works? Would we need to map out hundreds or thousands of separate circuits before anything useful could be understood? I didn't think so. History shows that the best solutions to scientific problems are simple and elegant. While the details may be forbidding and the road to a final theory may be arduous, the ultimate conceptual framework is generally simple.

Without a core explanation to guide inquiry, neuroscientists don't have much to go on as they try to assemble all the details they've collected into a coherent picture. The brain is incredibly complex, a vast and daunting tangle of cells. At first glance it looks like a stadium full of cooked spaghetti. It's also been described as an electrician's nightmare. But with close and careful inspection we see that the brain isn't a random heap. It has lots of organization and structure—but much too much of it for us to hope we'll be able to just intuit the workings of the whole, the way we're able to see how the shards of a broken vase fit back together. The failing isn't one of not having enough data or even the right pieces of data; what we need is a shift in perspective. With the proper framework, the details will become meaningful and manageable. Consider the following fanciful analogy to get the flavor of what I mean.

Imagine that millennia from now humans have gone extinct, and explorers from an advanced alien civilization land on Earth. They want to figure out how we lived. They are especially puzzled by our network of roadways. What were these bizarre elaborate structures for? They begin by cataloging everything, both via satellites and from the ground. They are meticulous archaeologists. They record the location of every stray fragment of asphalt, every signpost that has fallen over and been carried downhill by erosion, every detail they can find. They note that some road networks are different from others; in some places they are windy and narrow and almost random-looking, in some places they form a nice regular grid, and over some stretches they become thick and run for hundreds of miles through the desert. They collect a mountain of details, but these details don't mean anything to them. They continue to collect more detail in hopes of finding some new data that explain it all. They remain stumped for a long time.

That is, until one of them says, "Eureka! I think I see . . . these creatures couldn't teleport themselves like we can. They had to travel around from place to place, perhaps on mobile platforms of a cunning design." From this basic insight, many details begin to fall into place. The small twisty street networks are from early times when the means of conveyance were slow. The thick long roadways were for traveling long distances at high speeds, suggesting at last an explanation for why the signs on those roads had different numbers painted on them. The scientists start to deduce residential versus industrial zoning, the way the needs of commerce and transportation infrastructure might have interacted, and so on. Many of the details they had cataloged turn out to be not very relevant, just accidents of history or the requirements of local geography. The same amount of raw data exists, but it no longer is puzzling.

We can be confident that the same type of breakthrough will let us understand what all the brain's details are about.

Unfortunately, not everyone believes we can understand how the brain works. A surprising number of people, including a few neuroscientists, believe that somehow the brain and intelligence are beyond explanation. And some believe that even if we could understand them, it would be impossible to build machines that work the same way, that intelligence requires a human body, neurons, and perhaps some new and unfathomable laws of physics. Whenever I hear these arguments, I imagine the intellectuals of the past who argued against studying the heavens or fought against dissecting cadavers to see how our bodies worked. "Don't bother studying that, it will lead to no good, and even if you could understand how it works there is nothing we can do with that knowledge." Arguments like this one lead us to a branch of philosophy called functionalism, our last stop in this brief history of our thinking about thinking.

According to functionalism, being intelligent or having a mind is purely a property of organization and has nothing inherently to do with what you're organized out of. A mind exists in any system whose constituent parts have the right causal relationship with each other, but those parts can just as validly be neurons, silicon chips, or something else. Clearly, this view is standard issue to any would-be builder of intelligent machines.

Consider: Would a game of chess be any less real if it was played with a salt shaker standing in for a lost knight piece? Clearly not. The salt shaker is functionally equivalent to a "real" knight by virtue of how it moves on the board and interacts with the other pieces, so your game is truly a game of chess and not just a simulation of one. Or consider, wouldn't this sentence be the same if I were to go through it with my cursor deleting each character, then retyping it? Or to take an example closer to home, consider the fact that every few years your body replaces most of the atoms that comprise you. In spite of this, you remain yourself in all the ways that matter to you. One atom is as good as any other if it's playing the same functional role in your molecular makeup. The same story should hold for the brain: if a mad scientist were to replace each of your neurons with a functionally equivalent micromachine replica, you should come out of the procedure feeling no less your own true self than you had at the outset.

By this principle, an artificial system that used the same functional architecture as an intelligent, living brain should be likewise intelligent—and not just contrivedly so, but actually, truly intelligent.

AI proponents, connectionists, and I are all functionalists, insofar as we all believe there's nothing inherently special or magical about the brain that allows it to be intelligent. We all believe we'll be able to build intelligent machines, somehow, someday. But there are different interpretations of functionalism. While I've already stated what I consider the central failing of the AI and the connectionist paradigms—the input-output fallacy—there's a bit more worth saying about why we haven't yet been able to design intelligent machines. While the AI proponents take what I consider a self-defeating hard line, the connectionists, in my view, have mainly been just too timid.

AI researchers ask, "Why should we engineers be bound by the solutions evolution happened to stumble upon?" In principle, they have a point. Biological systems, like the brain and the genome, are viewed as notoriously inelegant. A common metaphor is that of the Rube Goldberg machine, named after the Depression-era cartoonist who drew comically overcomplicated contraptions to accomplish trivial tasks. Software designers have a related term, *kludge,* to refer to programs that are written without foresight and wind up full of burdensome, useless complexity, often to the point of becoming incomprehensible even to the programmers who wrote them. AI researchers fear the brain is similarly a mess, a several-hundred-million-year-old kludge, chock-full of inefficiencies and evolutionary "legacy code." If so, they wonder, why not just throw out the whole sorry clutter and start afresh?

Many philosophers and cognitive psychologists are sympathetic to this view. They love the metaphor of the mind being like software that's run by the brain, the organic analog of computer hardware. In computers, the hardware level and the software level are distinct from each other. The same software program can be made to run on any Universal Turing Machine. You can run WordPerfect on a PC, a Macintosh, or a Cray supercomputer, for example, even though all three systems have different hardware configurations. And the hardware has nothing of importance to teach you if you're trying to learn WordPerfect. By analogy, the thinking goes, the brain has nothing to teach us about the mind.

AI defenders also like to point out historical instances in which the engineering solution differs radically from nature's version. For example, how did we succeed in building flying machines? By imitating the flapping action of winged animals? No. We did it with fixed wings and propellers, and later with jet engines. It may not be how nature did it, but it works—and does so far better than flapping wings.

Similarly, we made a land vehicle that could outrun cheetahs not by making four-legged, cheetah-like running machines, but by inventing wheels. Wheels are a great way to move over flat terrain, and just because evolution never stumbled across that particular strategy doesn't mean it's not an excellent way for us to get around. Some philosophers of mind have taken a shine to the metaphor of the "cognitive wheel," that is, an AI solution to some problem that although entirely different from how the brain does it is just as good. In other words, a program that produces outputs resembling (or surpassing) human performance on a task in some narrow but useful way really is just as good as the way our brains do it.

I believe this kind of ends-justify-the-means interpretation of functionalism leads AI researchers astray. As Searle showed with the Chinese Room, behavioral equivalence is not enough. Since intelligence is an internal property of a brain, we have to look inside the brain to understand what intelligence is. In our investigations of the brain, and especially the neocortex, we will need to be careful in figuring out which details are just superfluous "frozen accidents" of our evolutionary past; undoubtedly, many Rube Goldberg–style processes are mixed in with the important features. But as we'll soon see, there is an underlying elegance of great power, one that surpasses our best computers, waiting to be extracted from these neural circuits.

Connectionists intuitively felt the brain wasn't a computer and that its secrets lie in how neurons behave when connected together. That was a good start, but the field barely moved on from its early successes. Although thousands of people worked on three-layer networks, and many still do, research on cortically realistic networks was, and remains, rare.

For half a century we've been bringing the full force of our species' considerable cleverness to trying to program intelligence into computers. In the process we've come up with word processors, databases, video games, the Internet, mobile phones, and convincing computer-animated dinosaurs. But intelligent machines still aren't anywhere in the picture. To succeed, we will need to crib heavily from nature's engine of intelligence, the neocortex. We have to extract intelligence from within the brain. No other road will get us there.

3

THE HUMAN BRAIN

So what makes the brain so unlike the programming that goes into AI and neural networks? What is so unusual about the brain's design, and why does it matter? As we'll see in the next few chapters, the brain's architecture has a great deal to tell us about how the brain really works and why it is fundamentally different from a computer.

Let's begin our introduction with the whole organ. Imagine there is a brain sitting on a table and we are dissecting it together. The first thing you'll notice is that the outer surface of the brain seems highly uniform. A pinkish gray, it resembles a smooth cauliflower with numerous ridges and valleys called gyri and sulci. It is soft and squishy to the touch. This is the neocortex, a thin sheet of neural tissue that envelops most of the older parts of the brain. We are going to focus most of our attention on the neocortex. Almost everything we think of as intelligence—perception, language, imagination, mathematics, art, music, and planning—occurs here. Your neocortex is reading this book.

Now, here I have to admit I am a neocortical chauvinist.

I know I'm going to meet some resistance on this point, so let me take a minute to defend my approach before we get too far in. Every part of the brain has its own community of scientists who study it, and the suggestion that we can get to the bottom of intelligence by understanding just the neocortex is sure to raise a few howls of objection from communities of offended researchers. They will say things like, "You cannot possibly understand the neocortex without understanding brain region *blah,* because the two are highly interconnected like so, and you need brain region *blah* to do such and such." I don't disagree. Granted, the brain consists of many parts and most of them are critical to being human. (Oddly, an exception is the part of the brain with the largest number of cells, the cerebellum. If you are born without a cerebellum or it is damaged, you can lead a pretty normal life. However, this is not true for most other brain regions; most are required for basic living, or sentience.)

My counterargument is that I am not interested in building humans. I want to understand intelligence and build intelligent machines. Being human and being intelligent are separate matters. An intelligent machine need not have sexual urges, hunger, a pulse, muscles, emotions, or a humanlike body. A human is much more than an intelligent machine. We are biological creatures with all the necessary and sometimes unwanted baggage that comes from eons of evolution. If you want to build intelligent machines that behave just like humans—that is, to pass the Turing Test in all ways—then you probably would have to recreate much of the other stuff that makes humans what we are. But as we will see later, to build machines that are undoubtedly intelligent but not exactly like humans, we can focus on the part of the brain strictly related to intelligence.

To those who may be offended by my singular focus on the neocortex, let me say I agree that other brain structures, such as the brain stem, basal ganglia, and amygdala, are indeed important to the functioning of the human neocortex. No question.

But I hope to convince you that all the essential aspects of intelligence occur in the neocortex, with important roles also played by two other brain regions, the thalamus and the hippocampus, that we will discuss later in the book. In the long run, we will need to understand the functional roles of all brain regions. But I believe those matters will be best addressed in the context of a good overall theory of neocortical function. That's my two cents on the matter. Let's get back to the neocortex, or its shorter moniker, the cortex.

Get six business cards or six playing cards—either will do—and put them in a stack. (It will really help if you do this instead of just imagining it.) You are now holding a model of the cortex. Your six business cards are about 2 millimeters thick and should give you a sense of how thin the cortical sheet is. Just like your stack of cards, the neocortex is about 2 millimeters thick and has six layers, each approximated by one card.

Stretched flat, the human neocortical sheet is roughly the size of a large dinner napkin. The cortical sheets of other mammals are smaller: the rat's is the size of a postage stamp; the monkey's is about the size of a business-letter envelope. But regardless of size, most of them contain six layers similar to what you see in your stack of business cards. Humans are smarter because our cortex, relative to body size, covers a larger area, not because our layers are thicker or contain some special class of "smart" cells. Its size is quite impressive as it surrounds and envelops most of the rest of the brain. To accommodate our large brain, nature had to modify our general anatomy. Human females developed a wide pelvis to give birth to big-headed children, a feature that some paleoanthropologists believe coevolved with the ability to walk on two legs. But that still wasn't enough, so evolution folded up the neocortex, stuffing it into our skulls like a sheet of paper crumpled into a brandy snifter.

Your neocortex is loaded with nerve cells, or neurons. They are so tightly packed that no one knows precisely how many cells it contains. If you draw a tiny square, one millimeter on a side (about half the size of this letter *o*), on the top of your stack of business cards, you are marking the position of an estimated one hundred thousand (100,000) neurons. Imagine trying to count the exact number in such a tiny space; it is virtually impossible. Nevertheless, some anatomists have estimated that the typical human neocortex contains around thirty billion neurons (30,000,000,000), but no one would be surprised if the figure was significantly higher or lower.

Those thirty billions cells are you. They contain almost all your memories, knowledge, skills, and accumulated life experience. After twenty-five years of thinking about brains, I still find this fact astounding. That a thin sheet of cells sees, feels, and creates our worldview is just short of incredible. The warmth of a summer day and the dreams we have for a better world are somehow the creation of these cells. Many years after he wrote his article in *Scientific American,* Francis Crick wrote a book about brains called *The Astonishing Hypothesis*. The astonishing hypothesis was simply that the mind is the creation of the cells in the brain. There is nothing else, no magic, no special sauce, only neurons and a dance of information. I hope you can get a sense of how incredible this realization is. There appears to be a large philosophical gulf between a collection of cells and our conscious experience, yet mind and brain are one and the same. In calling this a hypothesis, Crick was being politically correct. That the cells in our brains create the mind is a fact, not a hypothesis. We need to understand what these thirty billion cells do and how they do it. Fortunately, the cortex is not just an amorphous blob of cells. We can take a deeper look at its structure for ideas about how it gives rise to the human mind.

Let's go back to our dissection table and look at the brain some more. To the naked eye, the neocortex presents almost no landmarks. There are a few, to be sure, such as the giant fissure separating the two cerebral hemispheres and the prominent sulcus that divides the back and front regions. But just about everywhere you look, from left to right, from back to front, the convoluted surface looks pretty much the same. There are no visible boundary lines or color codes demarcating areas that specialize in different sensory information or different types of thought.

People have long known there are boundaries in there somewhere, though. Even before neuroscientists were able to discern anything helpful about the circuitry of the cortex, they knew that some mental functions were localized to certain regions of it. If a stroke knocks out Joe's right parietal lobe, he can lose his ability to perceive—or even conceive of—anything on the left side of his body or in the left half of space around himself. A stroke in the left frontal region known as Broca's area, by contrast, compromises his ability to use the rules of grammar, although his vocabulary and his ability to understand the meanings of words are unchanged. A stroke in an area called the fusiform gyrus can knock out the ability to recognize faces—Joe can't recognize his mother, his children, or even his own face in a photograph. Deeply fascinating disorders like these gave early neuroscientists the notion that the cortex consists of many functional regions or functional areas. The terms are equivalent.

We have learned a lot about functional areas in the past century, but much remains to be discovered. Each of these regions is semi-independent and seems to be specialized for certain aspects of perception or thought. Physically they are arranged in an irregular patchwork quilt, which varies only a little from person to person. Rarely are the functions cleanly delineated. Functionally they are arranged in a branching hierarchy.

The notion of a hierarchy is critical, so I want to take a moment to carefully define it. I'll be referring to it throughout the book. In a hierarchical system, some elements are in an abstract sense "above" and "below" others. In a business hierarchy, for example, a mid-level manager is above a mail clerk and below the vice president. This has nothing to do with physical aboveness or belowness; even if she works on a lower floor than the mail clerk, the manager is still "above" him hierarchically. I emphasize this point to make clear what I mean whenever I talk about one functional region being higher or lower than another. It has nothing to do with their physical arrangement in the brain. All the functional areas of the cortex reside in the same convoluted cortical sheet. What makes one region "higher" or "lower" than another is how they are connected to one another. In the cortex, lower areas feed information up to higher areas by way of a certain neural pattern of connectivity, while higher areas send feedback down to lower areas using a different connection pattern. There are also lateral connections between areas that are in separate branches of the hierarchy, like one mid-level manager communicating with her counterpart in a sister office in another state. A detailed map of the monkey cortex has been worked out by two scientists, Daniel Felleman and David van Essen. The map shows dozens of regions connected together in a complex hierarchy. We can assume the human cortex has a similar hierarchy.

The lowest of the functional regions, the primary sensory areas, are where sensory information first arrives in the cortex. These regions process the information at its rawest, most basic level. For example, visual information enters the cortex through the primary visual area, called V1 for short. V1 is concerned with low-level visual features such as tiny edge-segments, small-scale components of motion, binocular disparity (for stereovision), and basic color and contrast information. V1 feeds information up to other areas, such as V2, V4, and IT (we'll have more to say about them later), and to a bunch of other areas besides. Each of these areas is concerned with more specialized or abstract aspects of the information. For example, cells in V4 respond to objects of medium complexity such as star shapes in different colors like red or blue. Another area called MT specializes in the motions of objects. In the higher echelons of the visual cortex are areas that represent your visual memories of all sorts of objects like faces, animals, tools, body parts, and so on.

Your other senses have similar hierarchies. Your cortex has a primary auditory area called A1 and a hierarchy of auditory regions above it, and it has a primary somatosensory (body sense) area called S1 and a hierarchy of somatosensory regions above that. Eventually, sensory information passes into "association areas," which is the name sometimes used for the regions of the cortex that receive inputs from more than one sense. For example, your cortex has areas that receive input from both vision and touch. It is thanks to association regions that you are able to be aware that the sight of a fly crawling up your arm and the tickling sensation you feel there share the same cause. Most of these areas receive highly processed input from several senses, and their functions remain unclear. I will have much to say about the cortical hierarchy later in the book.

There is yet another set of areas in the frontal lobes of the brain that creates motor output. The motor system of the cortex is also hierarchically organized. The lowest area, M1, sends connections to the spinal cord and directly drives muscles. Higher areas feed sophisticated motor commands to M1. The hierarchy of the motor area and the hierarchies of sensory areas look remarkably similar. They seem to be put together in the same way. In the motor region we think of information flowing down the hierarchy toward M1 to drive the muscles and in the sensory regions we think of information flowing up the hierarchy away from the senses. But in reality information flows both ways. What is referred to as feedback in sensory regions is the output of the motor region, and vice versa.

Most descriptions of brains are based on flowcharts that reflect an oversimplified view of hierarchies. That is, input (sights, sounds, touch) flows into the primary sensory areas and gets processed as it moves up the hierarchy, then gets passed through the association areas, then gets passed to the frontal lobes of the cortex, and finally gets passed back down to the motor areas. I'm not saying this view is completely wrong. When you read aloud, visual information does indeed enter at V1, flows up to association areas, makes its way over to the frontal motor cortex, and winds up making the muscles in your mouth and throat form the sounds of speech. However, that isn't all there is to it. It's just not that simple. In the oversimplified view I am cautioning against, the process is generally treated as though information flows in a single direction, like widgets being built on a factory assembly line. But information in the cortex always flows in the opposite direction as well, and with many more projections feeding back down the hierarchy than up. As you read aloud, higher regions of your cortex send more signals "down" to your primary visual cortex than your eye receives from the printed page! We'll get to what those feedback projections are doing in later chapters. For now, I want to impress upon you one fact: although the up hierarchy is real, we have to be careful not to think that the information flow is all one way.

Back at the dissection table, let's assume we set up a powerful microscope, cut a thin slice from the cortical sheet, stain some of the cells, and take a look at our handiwork through the eyepiece. If we stained all the cells in our slice, we'd see a solid black mass because the cells are so tightly packed and intermingled. But if we use a stain that marks a smaller fraction of cells, we can see the six layers I mentioned. These layers are formed by variations in the density of cell bodies, cell types, and their connections.

All neurons have features in common. Apart from a cell body, which is the roundish part you imagine when you think of a cell, they also have branching, wirelike structures called axons and dendrites. When the axon from one neuron touches the dendrite of another, they form small connections called synapses. Synapses are where the nerve impulse from one cell influences the behavior of another cell. A neural signal, or spike, arriving at a synapse can make it more likely for the recipient cell to spike. Some synapses have the opposite effect, making it less likely the recipient cell will spike. Thus synapses can be inhibitory or excitatory. The strength of a synapse can change depending on the behavior of the two cells. The simplest form of this synaptic change is that when two neurons generate a spike at nearly the same time, the connection strength between the two neurons will be increased. I will say more about this process, called Hebbian learning, a bit later. In addition to changing the strength of a synapse, there is evidence that entirely new synapses can be formed between two neurons. This may be happening all the time, although the scientific evidence is controversial. Regardless of the details of how synapses change their strength, what is certain is that the formation and strengthening of synapses is what causes memories to be stored.

While there are many types of neurons in the neocortex, one broad class of them comprises eight out of every ten cells. These are the pyramidal neurons, so called because their cell bodies are shaped roughly like pyramids. Except for the top layer of the six-layered cortex, which has miles of axons but very few cells, every layer contains pyramidal cells. Each pyramidal neuron connects to many other neurons in its immediate neighborhood, and each sends a lengthy axon laterally out to more distant regions of cortex or down to lower brain structures like the thalamus.

A typical pyramidal cell has several thousand synapses. Again, it is very difficult to know exactly how many because of their extreme density and small size. The number of synapses varies from cell to cell, layer to layer, and region to region. If we were to take the conservative position that the average pyramidal cell has one thousand synapses (the actual number is probably close to five or ten thousand), then our neocortex would have roughly thirty trillion synapses altogether. That is an astronomically large number, well beyond our intuitive grasp. It is apparently sufficient to store all the things you can learn in a lifetime.

According to rumor, Albert Einstein once said that conceiving the theory of special relativity was straightforward, almost easy. It followed naturally from a single observation: that the speed of light is constant to all observers even if the observers are moving at different speeds. This is counterintuitive. It is like saying the speed of a thrown ball is always the same regardless of how hard it is thrown or how fast the individuals throwing and observing the ball are moving. Everybody sees the ball moving at the same speed relative to them under all circumstances. It doesn't seem like it could be true. But it was proven to be true for light, and Einstein cleverly asked what the consequences of this bizarre fact were. He methodically thought about all the implications of a constant speed of light, and he was led to the even more bizarre predictions of special relativity, such as time slowing down as you move faster, and energy and mass being fundamentally the same thing. Books on relativity walk through his line of reasoning with everyday examples of trains, bullets, flashlights, and so forth. The theory isn't hard, but it is definitely counterintuitive.

There is an analogous discovery in neuroscience—a fact about the cortex that is so surprising that some neuroscientists refuse to believe it and most of the rest ignore it because they don't know what to make of it. But it is a fact of such importance that if you carefully and methodically explore its implications, it will unravel the secrets of what the neocortex does and how it works. In this case, the surprising discovery came from the basic anatomy of the cortex itself, but it took an unusually insightful mind to recognize it. That person was Vernon Mountcastle, a neuroscientist at Johns Hopkins University in Baltimore. In 1978 he published a paper titled "An Organizing Principle for Cerebral Function." In this paper, Mountcastle points out that the neocortex is remarkably uniform in appearance and structure. The regions of cortex that handle auditory input look like the regions that handle touch, which look like the regions that control muscles, which look like Broca's language area, which look like practically every other region of the cortex. Mountcastle suggests that since these regions all look the same, perhaps they are actually performing the same basic operation! He proposes that the cortex uses the same computational tool to accomplish everything it does.

All anatomists at that time, and for decades prior to Mountcastle, recognized that the cortex looks similar everywhere; this is undeniable. But instead of asking what that could mean, they spent their time looking for the differences between one area of cortex and another. And they did find differences. They assumed that if one region is used for language and another for vision, then there ought to be differences between those regions. If you look closely enough you find them. Regions of cortex vary in thickness, cell density, relative proportion of cell types, length of horizontal connections, synapse density, and many other ways that can be tricky to discover. One of the most-studied regions, the primary visual area V1, actually has a few extra divisions in one of its layers. The situation is analogous to the work of biologists in the 1800s. They spent their time discovering the minute differences between species. Success for them was finding that two mice that looked nearly identical were actually separate species. For many years Darwin followed the same course, often studying mollusks. But Darwin eventually had the big insight to ask how all these species could be so similar. It is their similarity that is surprising and interesting, much more so than their differences.

Mountcastle makes a similar observation. In a field of anatomists looking for minute differences in cortical regions, he shows that despite the differences, the neocortex is remarkably uniform. The same layers, cell types, and connections exist throughout. It looks like the six business cards everywhere. The differences are often so subtle that trained anatomists can't agree on them. Therefore, Mountcastle argues, all regions of the cortex are performing the same operation. The thing that makes the vision area visual and the motor area motoric is how the regions of cortex are connected to each other and to other parts of the central nervous system.

In fact, Mountcastle argues that the reason one region of cortex looks slightly different from another is because of what it is connected to, and not because its basic function is different. He concludes that there is a common function, a common algorithm, that is performed by all the cortical regions. Vision is no different from hearing, which is no different from motor output. He allows that our genes specify how the regions of cortex are connected, which is very specific to function and species, but the cortical tissue itself is doing the same thing everywhere.

Let's think about this for a moment. To me, sight, hearing, and touch seem very different. They have fundamentally different qualities. Sight involves color, texture, shape, depth, and form. Hearing has pitch, rhythm, and timbre. They feel very different. How can they be the same? Mountcastle says they aren't the same, but the way the cortex processes signals from the ear is the same as the way it processes signals from the eyes. He goes on to say that motor control works on the same principle, too.

Scientists and engineers have for the most part been ignorant of, or have chosen to ignore, Mountcastle's proposal. When they try to understand vision or make a computer that can "see," they devise vocabulary and techniques specific to vision. They talk about edges, textures, and three-dimensional representations. If they want to understand spoken language, they build algorithms based on rules of grammar, syntax, and semantics. But if Mountcastle is correct, these approaches are not how the brain solves these problems, and are therefore likely to fail. If Mountcastle is correct, the algorithm of the cortex must be expressed independently of any particular function or sense. The brain uses the same process to see as to hear. The cortex does something universal that can be applied to any type of sensory or motor system.

When I first read Mountcastle's paper I nearly fell out of my chair. Here was the Rosetta stone of neuroscience—a single paper and a single idea that united all the diverse and wondrous capabilities of the human mind. It united them under a single algorithm. In one step it exposed the fallacy of all previous attempts to understand and engineer human behavior as diverse capabilities. I hope you can appreciate how radical and wonderfully elegant Mountcastle's proposal is. The best ideas in science are always simple, elegant, and unexpected, and this is one of the best. In my opinion it was, is, and will likely remain the most important discovery in neuroscience. Incredibly, though, most scientists and engineers either refuse to believe it, choose to ignore it, or aren't aware of it.

Part of this neglect stems from a poverty of tools for studying how information flows within the six-layered cortex. The tools we do have operate on a grosser level and are generally aimed at locating where in the cortex, as opposed to when and how, various capabilities arise. For example, much of the neuroscience reported in the popular press these days implicitly favors the idea of the brain as a collection of highly specialized modules. Functional imaging techniques like functional MRI and PET scanning focus almost exclusively on brain maps and the functional regions I mentioned earlier. Typically in these experiments, a volunteer subject lies down with his or her head inside the scanner and performs some kind of mental or motor task. It might be playing a video game, generating verb conjugations, reading sentences, looking at faces, naming pictures, imagining something, memorizing lists, making financial decisions, and so on. The scanner detects which brain regions are more active than usual during these tasks and draws colored splotches over an image of the subject's brain to pinpoint them. These regions are presumably central to the task. Thousands of functional imaging experiments have been done and thousands more will follow. Through the course of it all, we are gradually building up a picture of where certain functions happen in the typical adult brain. It is easy to say, "this is the face recognition area, this is the math area, this is the music area," and so on. Since we don't know how the brain accomplishes these tasks, it is natural to assume that the brain carries out the various activities in different ways.

But does it? A growing and fascinating body of evidence supports Mountcastle's proposal. Some of the best examples demonstrate the extreme flexibility of the neocortex. Any human brain, if nourished properly and put in the right environment, can learn any of thousands of spoken languages. That same brain can also learn sign language, written language, musical language, mathematical language, computer languages, and body language. It can learn to live in frigid northern climes or in a scorching desert. It can become an expert in chess, fishing, farming, or theoretical physics. Consider the fact that you have a special visual area that seems to be specifically devoted to representing written letters and digits. Does this mean you were born with a language area ready to process letters and digits? Unlikely. Written language is far too recent an invention for our genes to have evolved a specific mechanism for it. So the cortex is still dividing itself into task-specific functional areas long into childhood, based purely on experience. The human brain has an incredible capacity to learn and adapt to thousands of environments that didn't exist until very recently. This argues for an extremely flexible system, not one with a thousand solutions for a thousand problems.

Neuroscientists have also found that the wiring of the neocortex is amazingly "plastic," meaning it can change and rewire itself depending on the type of inputs flowing into it. For example, newborn ferret brains can be surgically rewired so that the animals' eyes send their signals to the areas of cortex where hearing normally develops. The surprising result is that the ferrets develop functioning visual pathways in the auditory portions of their brains. In other words, they see with brain tissue that normally hears sounds. Similar experiments have been done with other senses and brain regions. For instance, pieces of rat visual cortex can be transplanted around the time of birth to regions where the sense of touch is usually represented. As the rat matures, the transplanted tissue processes touch rather than vision. Cells were not born to specialize in vision or touch or hearing.

Human neocortex is every bit as plastic. Adults who are born deaf process visual information in areas that normally become auditory regions. And congenitally blind adults use the rearmost portion of their cortex, which ordinarily becomes dedicated to vision, to read braille. Since braille involves touch, you might think it would primarily activate touch regions—but apparently no area of cortex is content to represent nothing. The visual cortex, not receiving information from the eyes like it is "supposed" to, casts around for other input patterns to sift through—in this case, from other cortical regions.

All this goes to show that brain regions develop specialized functions based largely on the kind of information that flows into them during development. The cortex is not rigidly designed to perform different functions using different algorithms any more than the earth's surface was predestined to end up with its modern arrangement of nations. The organization of your cortex, like the political geography of the globe, could have turned out differently given a different set of early circumstances.

Genes dictate the overall architecture of the cortex, including the specifics of what regions are connected together, but within that structure the system is highly flexible.

Mountcastle was right. There is a single powerful algorithm implemented by every region of cortex. If you connect regions of cortex together in a suitable hierarchy and provide a stream of input, it will learn about its environment. Therefore, there is no reason for intelligent machines of the future to have the same senses or capabilities as we humans. The cortical algorithm can be deployed in novel ways, with novel senses, in a machined cortical sheet so that genuine, flexible intelligence emerges outside of biological brains.

Let's move on to a topic that is related to Mountcastle's proposal and is equally surprising. The inputs to your cortex are all basically alike. Again, you probably think of your senses as being completely separate entities. After all, sound is carried as compression waves through air, vision is carried as light, and touch is carried as pressure on your skin. Sound seems temporal, vision seems mainly pictorial, and touch seems essentially spatial. What could be more different than the sound of a bleating goat versus the sight of an apple versus the feel of a baseball?

But let's take a closer look. Visual information from the outside world is sent to your brain via a million fibers in your optic nerve. After a brief transit through the thalamus, they arrive at the primary visual cortex. Sounds are carried in via the thirty thousand fibers of your auditory nerve. They pass through some older parts of your brain and then arrive at your primary auditory cortex. Your spinal cord carries information about touch and internal sensations to your brain via another million fibers. They are received by your primary somatosensory cortex. These are the main inputs to your brain. They are how you sense the world.

You can visualize these inputs as a bundle of electrical wires or a bundle of optical fibers. You might have seen lamps made with optical fibers where pinpoints of colored light appear at the end of each fiber. The inputs to the brain are like this, but the fibers are called axons, and they carry neural signals called "action potentials" or "spikes," which are partly chemical and partly electrical. The sense organs supplying these signals are different, but once they are turned into brain-bound action potentials, they are all the same—just patterns.

If you look at a dog, for example, a set of patterns will flow through the fibers of your optic nerve into the visual part of your cortex. If you listen to the dog bark, a different set of patterns will flow along your auditory nerve and into the hearing parts of your brain. If you pet the dog, a set of touch-sensation patterns will flow from your hand, through fibers in your spine, and into the parts of your brain that deal with touch. Each pattern—see the dog, hear the dog, feel the dog—is experienced differently because each gets channeled through a different path in the cortical hierarchy. It matters where the cables go to inside the brain. But at the abstract level of sensory inputs, these are all essentially the same, and are all handled in similar ways by the six-layered cortex. You hear sound, see light, and feel pressure, but inside your brain there isn't any fundamental difference between these types of information. An action potential is an action potential. These momentary spikes are identical regardless of what originally caused them. All your brain knows is patterns.

Your perceptions and knowledge about the world are built from these patterns. There's no light inside your head. It's dark in there. There's no sound entering your brain either. It's quiet inside. In fact, the brain is the only part of your body that has no senses itself. A surgeon could stick a finger into your brain and you wouldn't feel it. All the information that enters your mind comes in as spatial and temporal patterns on the axons.

What exactly do I mean by spatial and temporal patterns? Let's look at each of our main senses in turn. Vision carries both spatial and temporal information. *Spatial patterns* are coincident patterns in time; they are created when multiple receptors in the same sense organ are stimulated simultaneously. In vision, the sense organ is your retina. An image enters your pupil, gets inverted by your lens, hits your retina, and creates a spatial pattern. This pattern gets relayed to your brain. People tend to think that there's a little upside-down picture of the world going into your visual areas, but that's not how it works. There is no picture. It's not an image anymore. Fundamentally, it is just electrical activity firing in patterns. Its imagelike qualities get lost very rapidly as your cortex handles the information, passing components of the pattern up and down between different areas, sifting them, filtering them.

Vision also relies on *temporal patterns,* which means the patterns entering your eyes are constantly changing over time. But while the spatial aspect of vision is intuitively obvious, its temporal aspect is less apparent. About three times every second, your eyes make a sudden movement called a saccade. They fixate on one point, and then suddenly jump to another point. Every time your eyes move, the image on your retina changes. This means that the patterns carried into your brain are also changing completely with each saccade. And that's in the simplest possible case of you just sitting still looking at an unchanging scene. In real life, you constantly move your head and body and walk through continuously shifting environments. Your conscious impression is of a stable world full of objects and people that are easy to keep track of. But this impression is only made possible by your brain's ability to deal with a torrent of retinal images that never repeat a pattern exactly. Natural vision, experienced as patterns entering the brain, flows like a river. Vision is more like a song than a painting.

Many vision researchers ignore saccades and the rapidly changing patterns of vision. Working with anesthetized animals, they study how vision occurs when an unconscious animal fixates on a point. In doing so, they're taking away the time dimension. There's nothing wrong with that in principle; eliminating variables is a core element of the scientific method. But they're throwing away a central component of vision, what it actually consists of. Time needs a central place in a neuroscientific account of vision.

With hearing, we're used to thinking about sound's temporal aspect. It is intuitively obvious to us that sounds, spoken language, and music change over time. You can't listen to a song all at once any more than you can hear a spoken sentence instantaneously. A song only exists over time. Therefore we don't usually think of sound as a spatial pattern. In a way, it's the inverse of the case with vision: the temporal aspect is immediately apparent, but its spatial aspect is less obvious.

Hearing has a spatial component as well. You convert sounds into action potentials through a coiled-up organ in each ear called the cochlea. Tiny, opaque, spiral-shaped, and embedded in the hardest bone in the body, the temporal bone, the cochlea was deciphered more than half a century ago by a Hungarian physicist, Georg von Beksey. Building models of the inner ear, von Beksey discovered that each component of sound you hear causes a different portion of the cochlea to vibrate. High-frequency tones cause vibrations in the cochlea's stiff base. Low-frequency tones cause vibrations in the cochlea's floppier and wider outer portion. Mid-frequency tones vibrate intermediate segments. Each site on the cochlea is studded with neurons that fire as they are shaken. In daily life your cochleas are being vibrated by large numbers of simultaneous frequencies all the time. So each moment there is a new spatial pattern of stimulation along the length of each cochlea. Each moment a new spatial pattern streams up the auditory nerve. Again we see that this sensory information boils down to spatial-temporal patterns.

People don't usually think of touch as a temporal phenomenon, but it is every bit as time-based as it is spatial. You can carry out an experiment to see for yourself. Ask a friend to cup his hand, palm face up, and close his eyes. Place a small ordinary object in his palm—a ring, an eraser, anything will do—and ask him to identify it without moving any part of his hand. He won't have a clue other than weight and maybe gross size. Then tell him to keep his eyes closed and move his fingers over the object. He'll most likely identify it at once. By allowing the fingers to move, you've added time to the sensory perception of touch. There's a direct analogy between the fovea at the center of your retina and your fingertips, both of which have high acuity. So touch, too, is like a song. Your ability to make complex use of touch, such as buttoning your shirt or unlocking your front door in the dark, depends on continuous time-varying patterns of touch sensation.

We teach our children that humans have five senses: sight, hearing, touch, smell, and taste. We really have more. Vision is more like three senses—motion, color, and luminance (black-and-white contrast). Touch has pressure, temperature, pain, and vibration. We also have an entire system of sensors that tell us about our joint angles and bodily position. It is called the proprioceptive system (*proprio-* has the same Latin root as *proprietary* and *property*). You couldn't move without it. We also have the vestibular system in the inner ear, which gives us our sense of balance. Some of these senses are richer and more apparent to us than others, but they all enter our brain as streams of spatial patterns flowing through time on axons.

Your cortex doesn't really know or sense the world directly.

The only thing the cortex knows is the pattern streaming in on the input axons. Your perceived view of the world is created from these patterns, including your sense of self. In fact, your brain can't directly know where your body ends and the world begins. Neuroscientists studying body image have found that our sense of self is a lot more flexible than it feels. For example, if I give you a little rake and have you use it for reaching and grasping instead of using your hand, you will soon feel that it has become a part of your body. Your brain will change its expectations to accommodate the new patterns of tactile input. The rake is literally incorporated into your body map.

The idea that patterns from different senses are equivalent inside your brain is quite surprising, and although well understood, it still isn't widely appreciated. More examples are in order. The first one you can reproduce at home. All you need is a friend, a freestanding cardboard screen, and a fake hand. For your first time running this experiment, it would be ideal if you had a rubber hand, such as you might buy at a Halloween store, but it will also work if you just trace your hand on a sheet of blank paper. Lay your real hand on a tabletop a few inches away from the fake one and align them the same (fingertips pointed in the same direction, palms either both up or both down). Then place the screen between the two hands so that all you can see is the false one. While you stare at the fake hand, your friend's job is to simultaneously stroke both hands at corresponding points. For example, your friend could stroke both pinkies from knuckle to nail at the same speed, then issue three quick taps to the second joint of both index fingers with the same timing, then stroke a few light circles on the back of each hand, and so on. After a short time, areas in your brain where visual and somatosensory patterns come together—one of those association areas I mentioned earlier in this chapter—become confused.

You will actually feel the sensations being applied to the dummy hand as if it were your own.

Another fascinating example of this "pattern equivalency" is called sensory substitution. It may revolutionize life for people who lose their sight in childhood, and might someday be a boon to people who are born blind. It also might spawn new machine interface technologies for the rest of us.

Realizing that the brain is all about patterns, Paul Bach y Rita, a professor of biomedical engineering at the University of Wisconsin, has developed a method for displaying visual patterns on the human tongue. Wearing this display device, blind persons are learning to "see" via sensations on the tongue.

Here is how it works. The subject wears a small camera on his forehead and a chip on his tongue. Visual images are translated pixel for pixel into points of pressure on the tongue. A visual scene that can be displayed as hundreds of pixels on a crude television screen can be turned into a pattern of hundreds of tiny pressure points on the tongue. The brain quickly learns to interpret the patterns correctly.

One of the first people to wear the tongue-mounted device is Erik Weihenmayer, a world-class athlete who went blind at age thirteen and who lectures widely about not letting blindness stop his ambitions. In 2002, Weihenmayer summited Mount Everest, becoming the first blind person ever to undertake, much less accomplish, such a goal.

In 2003, Weihenmayer tried on the tongue unit and saw images for the first time since his childhood. He was able to discern a ball rolling on the floor toward him, reach for a soft drink on a table, and play the game Rock, Paper, Scissors. Later he walked down a hallway, saw the door openings, examined a door and its frame, and noted that there was a sign on it. Images initially experienced as sensations on the tongue were soon experienced as images in space.

These examples show once again that the cortex is extremely

flexible and that the inputs to the brain are just patterns. It doesn't matter where the patterns come from; as long as they correlate over time in consistent ways, the brain can make sense of them.

All of this shouldn't be too surprising if we take the view that patterns are all the brain knows about. Brains are pattern machines. It's not incorrect to express the brain's functions in terms of hearing or vision, but at the most fundamental level, patterns are the name of the game. No matter how different the activities of various cortical areas may seem from each other, the same basic cortical algorithm is at work. The cortex doesn't care if the patterns originated in vision, hearing, or another sense. It doesn't care if its inputs are from a single sensory organ or from four. Nor would it care if you happened to perceive the world with sonar, radar, or magnetic fields, or if you had tentacles rather than hands, or even if you lived in a world of four dimensions rather than three.

This means you don't need any one of your senses or any particular combination of senses to be intelligent. Helen Keller had no sight and no hearing, yet she learned language and became a more skillful writer than most sighted and hearing people. Here was a very intelligent person without two of our main senses, yet the incredible flexibility of the brain allowed her to perceive and understand the world as individuals with all five senses do.

This kind of remarkable flexibility in the human mind gives me high hopes for the brain-inspired technology we will create. When I think about building intelligent machines, I wonder, Why stick to our familiar senses? As long as we can decipher the neocortical algorithm and come up with a science of patterns, we can apply it to any system that we want to make intelligent. And one of the great features of neocortically inspired circuitry is that we won't need to be especially clever in programming it.

Just as auditory cortex can become "visual" cortex in a rewired ferret, just as visual cortex finds alternative usage in blind people, a system running the neocortical algorithm will be intelligent based on whatever kinds of patterns we choose to give it. We will still need to be smart about setting up the broad parameters of the system, and we will need to train and educate it. But the billions of neural details involved in the brain's ability to have complex, creative thoughts will take care of themselves, as naturally as they do in our children.

Finally, the idea that patterns are the fundamental currency of intelligence leads to some interesting philosophical questions. When I sit in a room with my friends, how do I know they are there or even if they are real? My brain receives a set of patterns that are consistent with patterns I have experienced in the past. These patterns correspond to people I know, their faces, their voices, how they usually behave, and all kinds of facts about them. I have learned to expect these patterns to occur together in predictable ways. But when you come down to it, it's all just a model. All our knowledge of the world is a model based on patterns. Are we certain the world is real? It's fun and odd to think about. Several science-fiction books and movies explore this theme. This is not to say that the people or objects aren't really there. They are really there. But our certainty of the world's existence is based on the consistency of patterns and how we interpret them. There is no such thing as direct perception. We don't have a "people" sensor. Remember, the brain is in a dark quiet box with no knowledge of anything other than the time-flowing patterns on its input fibers. Your perception of the world is created from these patterns, nothing else. Existence may be objective, but the spatial-temporal patterns flowing into the axon bundles in our brains are all we have to go on.

This discussion highlights the sometimes-questioned relationship between hallucination and reality. If you can hallucinate sensations coming from a rubber hand and you can "see" via touch stimulation of your tongue, are you being equally "fooled" when you sense touch on your own hand or see with your eyes? Can we trust that the world is as it seems? Yes. The world really does exist in an absolute form very close to how we perceive it. However, our brains can't know about the absolute world directly.

The brain knows about the world through a set of senses, which can only detect parts of the absolute world. The senses create patterns that are sent to the cortex, and processed by the same cortical algorithm to create a model of the world. In this way, spoken language and written language are perceived remarkably similarly, despite being completely different at the sensory level. Likewise, Helen Keller's model of the world was very close to yours and mine, despite the fact that she had a greatly reduced set of senses. Through these patterns the cortex constructs a model of the world that is close to the real thing, and then, remarkably, holds it in memory. It is memory—what happens to those patterns after they enter the cortex—that we'll discuss in the next chapter.

4

MEMORY

As you read this book, walk down a crowded street, hear a symphony, or comfort a crying child, your brain is being flooded with the spatial and temporal patterns from all of your senses. The world is an ocean of constantly changing patterns that come lapping and crashing into your brain. How do you manage to make sense of the onslaught? Patterns stream in, pass through various parts of the old brain, and eventually arrive at the neocortex. But what happens to them when they enter the cortex?

From the dawn of the industrial revolution, people have viewed the brain as some sort of machine. They knew there weren't gears and cogs in the head, but it was the best metaphor they had. Somehow information entered the brain and the brain-machine determined how the body should react. During the computer age, the brain has been viewed as a particular type of machine, the programmable computer. And as we saw in chapter 1, AI researchers have stuck with this view, arguing that their lack of progress is only due to how small and slow computers remain compared to the human brain. Today's computers may be equivalent only to a cockroach brain, they say, but when we make bigger and faster computers they will be as intelligent as humans.

There is a largely ignored problem with this brain-as-computer analogy. Neurons are quite slow compared to the transistors in a computer. A neuron collects inputs from its synapses, and combines these inputs together to decide when to output a spike to other neurons. A typical neuron can do this and reset itself in about five milliseconds (5 ms), or around two hundred times per second. This may seem fast, but a modern silicon-based computer can do one billion operations in a second. This means a basic computer operation is five million times faster than the basic operation in your brain! That is a very, very big difference. So how is it possible that a brain could be faster and more powerful than our fastest digital computers? "No problem," say the brain-as-computer people. "The brain is a parallel computer. It has billions of cells all computing at the same time. This parallelism vastly multiplies the processing power of the biological brain."

I always felt this argument was a fallacy, and a simple thought experiment shows why. It is called the "one hundred–step rule." A human can perform significant tasks in much less time than a second. For example, I could show you a photograph and ask you to determine if there is cat in the image. Your job would be to push a button if there is a cat, but not if you see a bear or a warthog or a turnip. This task is difficult or impossible for a computer to perform today, yet a human can do it reliably in half a second or less. But neurons are slow, so in that half a second, the information entering your brain can only traverse a chain one hundred neurons long. That is, the brain "computes" solutions to problems like this in one hundred steps or fewer, regardless of how many total neurons might be involved. From the time light enters your eye to the time you press the button, a chain no longer than one hundred neurons could be involved. A digital computer attempting to solve the same problem would take billions of steps. One hundred computer instructions are barely enough to move a single character on the computer's display, let alone do something interesting.

But if I have many millions of neurons working together, isn't that like a parallel computer? Not really. Brains operate in parallel and parallel computers operate in parallel, but that's the only thing they have in common. Parallel computers combine many fast computers to work on large problems such as computing tomorrow's weather. To predict the weather you have to compute the physical conditions at many points on the planet. Each computer can work on a different location at the same time. But even though there may be hundreds or even thousands of computers working in parallel, the individual computers still need to perform billions or trillions of steps to accomplish their task. The largest conceivable parallel computer can't do anything useful in one hundred steps, no matter how large or how fast.

Here is an analogy. Suppose I ask you to carry one hundred stone blocks across a desert. You can carry one stone at a time and it takes a million steps to cross the desert. You figure this will take a long time to complete by yourself, so you recruit a hundred workers to do it in parallel. The task now goes a hundred times faster, but it still requires a minimum of a million steps to cross the desert. Hiring more workers—even a thousand workers—wouldn't provide any additional gain. No matter how many workers you hire, the problem cannot be solved in less time than it takes to walk a million steps. The same is true for parallel computers. After a point, adding more processors doesn't make a difference. A computer, no matter how many processors it might have and no matter how fast it runs, cannot "compute" the answer to difficult problems in one hundred steps.

So how can a brain perform difficult tasks in one hundred steps that the largest parallel computer imaginable can't solve in a million or a billion steps? The answer is the brain doesn't "compute" the answers to problems; it retrieves the answers from memory. In essence, the answers were stored in memory a long time ago. It only takes a few steps to retrieve something from memory. Slow neurons are not only fast enough to do this, but they constitute the memory themselves. The entire cortex is a memory system. It isn't a computer at all.

Let me show, through an example, the difference between *computing* a solution to a problem and *using memory* to solve the same problem. Consider the task of catching a ball. Someone throws a ball to you, you see it traveling toward you, and in less than a second you snatch it out of the air. This doesn't seem too difficult—until you try to program a robot arm to do the same. As many a graduate student has found out the hard way, it seems nearly impossible. When engineers or computer scientists tackle this problem, they first try to calculate the flight of the ball to determine where it will be when it reaches the arm. This calculation requires solving a set of equations of the type you learn in high school physics. Next, all the joints of a robotic arm have to be adjusted in concert to move the hand into the proper position. This involves solving another set of mathematical equations more difficult than the first. Finally, this whole operation has to be repeated multiple times, for as the ball approaches, the robot gets better information about the ball's location and trajectory. If the robot waits to start moving until it knows exactly where the ball will arrive it will be too late to catch it. It has to start moving to catch the ball when it has only a poor sense of its location and it continually adjusts as the ball gets closer. A computer requires millions of steps to solve the numerous mathematical equations to catch the ball. And although a computer might be programmed to successfully solve this problem, the one hundred-step rule tells us that a brain solves it in a different way. It uses memory.

How do you catch the ball using memory? Your brain has a stored memory of the muscle commands required to catch a ball (along with many other learned behaviors). When a ball is thrown, three things happen. First, the appropriate memory is automatically recalled by the sight of the ball. Second, the memory actually recalls a temporal sequence of muscle commands. And third, the retrieved memory is adjusted as it is recalled to accommodate the particulars of the moment, such as the ball's actual path and the position of your body. The memory of how to catch a ball was not programmed into your brain; it was learned over years of repetitive practice, and it is stored, not calculated, in your neurons.

You might be thinking, "Wait a minute. Each catch is slightly different. You just said the recalled memory gets continually adjusted to accommodate the variations of where the ball is on any particular throw . . . Doesn't that require solving the same equations we were trying to avoid?" It may seem so, but nature solved the problem of variation in a different and very clever way. As we'll see later in this chapter, the cortex creates what are called *invariant representations,* which handle variations in the world automatically. A helpful analogy might be to imagine what happens when you sit down on a water bed: the pillows and any other people on the bed are all spontaneously pushed into a new configuration. The bed doesn't compute how high each object should be elevated; the physical properties of the water and the mattress's plastic skin take care of the adjustment automatically. As we'll see in the next chapter, the design of the six-layered cortex does something similar, loosely speaking, with the information that flows through it.

So the neocortex is not like a computer, parallel or otherwise. Instead of computing answers to problems the neocortex uses stored memories to solve problems and produce behavior.

Computers have memory too, in the form of hard drives and memory chips; however, there are four attributes of neocortical memory that are fundamentally different from computer memory:

- **The neocortex stores sequences of patterns.**
- **The neocortex recalls patterns auto-associatively.**
- **The neocortex stores patterns in an invariant form.**
- **The neocortex stores patterns in a hierarchy.**

We will discuss the first three differences in this chapter. I introduced the concept of hierarchy in the neocortex in chapter 3. In chapter 6, I will describe its significance and how it works.

The next time you tell a story, step back and consider how you can only relate one aspect of the tale at a time. You cannot tell me everything that happened all at once, no matter how quickly you talk or I listen. You need to finish one part of the story before you can move on to the next. This isn't only because spoken language is serial; written, oral, and visual storytelling all convey a narrative in a serial fashion. It is because the story is stored in your head in a sequential fashion and can only be recalled in the same sequence. You can't remember the entire story at once. In fact, it's almost impossible to think of anything complex that isn't a series of events or thoughts.

You may have noticed, too, that in telling a story some people can't get to the crux of it right away. They seem to ramble on with irrelevant details and tangents. This can be irritating. You want to scream, "Get to the point!" But they are chronicling the story as it happened to them, through time, and cannot tell it any other way.

Another example: I'd like you to imagine your home right now. Close your eyes and visualize it. In your imagination, go to the front door. Imagine what it looks like. Open your front door. Move inside. Now look to your left. What do you see? Look to the right. What is there? Go to your bathroom. What's on the right? What's on the left? What's in the top right drawer? What items do you keep in your shower? You know all these things plus thousands more and can recall them in great detail. These memories are stored in your cortex. You might say these things are all part of the memory of your home. But you can't think of them all at once. They are obviously related memories but there is no way you can bring to mind all of this detail at once. You have a thorough memory of your home; but to recall it you have to go through it in sequential segments, in much the same way as you experience it.

All memories are like this. You have to walk through the temporal sequence of how you do things. One pattern (approach the door) evokes the next pattern (go through the door), which evokes the next pattern (either go down the hall or ascend the stairs), and so on. Each is a sequence you've followed before. Of course, with a conscious effort I can change the order of how I describe my home to you. I can jump from basement to the second floor if I decide to focus on items in a nonsequential way. Yet once I start to describe any room or item I've chosen, I'm back to following a sequence. Truly random thoughts don't exist. Memory recall almost always follows a pathway of association.

You know the alphabet. Try saying it backward. You can't because you don't usually experience it backward. If you want to know what it's like to be a child learning the alphabet, try saying it in reverse. That's exactly what they're confronted with. It's really hard. Your memory of the alphabet is a sequence of patterns. It isn't something stored or recalled in an instant or in an arbitrary order. The same thing goes for the days of the week, the months of the year, your phone number, and countless other things.

Your memory for songs is a great example of temporal sequences in memory. Think of a tune you know. I like to use

"Somewhere over the Rainbow," but any melody will suffice. You cannot imagine the entire song at once, only in sequence. You can start at the beginning or maybe with the chorus, and then you play through it, filling in the notes one after another. You can't recall the song backward, just as you can't recall it all at once. You were first exposed to "Somewhere over the Rainbow" as it played through time, and you can only recall it in the same way you learned it.

This applies to very low level sensory memories too. Consider your tactile memory for textures. Your cortex has memories of what it feels like to hold a fistful of gravel, slide your fingers over velvet, and press down on a piano key. These memories are based on sequences every bit as much as the alphabet and songs are; it's just that the sequences are shorter, spanning mere fractions of a second rather than many seconds or minutes. If I buried your hand in a bucket of gravel while you slept, when you woke up you wouldn't know what you were touching until you moved your fingers. Your memory for the tactile texture of gravel is based on pattern sequences across the pressure- and vibration-sensing neurons in your skin. These sequences are different from those you'd receive if your hand was buried in sand or Styrofoam pellets or dry leaves. As soon as you flexed your hand, the scraping and rolling of the pebbles would create the telltale pattern sequences of gravel and trigger the appropriate memory in your somatosensory cortex.

The next time you get out of the shower, pay attention to how you dry yourself off with a towel. I discovered that I dry myself off with nearly the exact same sequence of rubs, pats, and body positions each time. And via a pleasant experiment I discovered that my wife also follows a semirigid pattern when she steps out of the shower. You probably do too. If you follow a sequence, try changing it. You can will yourself to do it, but you need to stay focused. If your attention wanders, you'll fall back into your accustomed pattern.

All memories are stored in the synaptic connections between neurons. Given the very large number of things we have stored in our cortex, and that at any moment in time we can recall only a tiny fraction of these stored memories, it stands to reason that only a limited number of synapses and neurons in your brain are playing an active role in memory recall at any one time. As you start to recall what is in your home, one set of neurons becomes active, which then leads to another set of neurons being active, and so on. An adult human neocortex has an incredibly large memory capacity. But, even though we have stored so many things, we can only remember a few at any time and can only do so in a sequence of associations.

Here is a fun exercise. Try to recall details from your past, details of where you lived, places you visited, and people you knew. I find I can always uncover memories of things I haven't thought of in many years. There are thousands of detailed memories stored in the synapses of our brains that are rarely used. At any point in time we recall only a tiny fraction of what we know. Most of the information is sitting there idly waiting for the appropriate cues to invoke it.

Computer memory does not normally store sequences of patterns. It can be made to do so using various software tricks (such as when you store a song on your computer), but computer memory does not do this automatically. In contrast, the cortex does store sequences automatically. Doing so is an inherent aspect of the neocortical memory system.

Now let's consider the second key feature of our memory, its auto-associative nature. As we saw in chapter 2, the term simply means that patterns are associated with themselves. An auto-associative memory system is one that can recall complete patterns when given only partial or distorted inputs. This can work for both spatial and temporal patterns. If you see your child's shoes sticking out from behind the draperies, you automatically envision his or her entire form. You complete the spatial pattern from a partial version of it. Or imagine you see a person waiting for a bus but can only see part of her because she is standing partially behind a bush. Your brain is not confused. Your eyes only see parts of a body, but your brain fills in the rest, creating a perception of a whole person that's so strong you may not even realize you're only inferring.

You also complete temporal patterns. If you recall a small detail about something that happened long ago, the entire memory sequence can come flooding back into your mind. Marcel Proust's famous series of novels, *Remembrance of Things Past,* opened with the memory of how a madeleine cookie smelled—and he was off and running for a thousand-plus pages. During conversation we often can't hear all the words if we are in a noisy environment. No problem. Our brains fill in what they miss with what they expect to hear. It's well established that we don't actually hear all the words we perceive. Some people complete others' sentences aloud, but in our minds all of us are doing this constantly. And not just the ends of sentences, but the middles and beginnings as well. For the most part we are not aware that we're constantly completing patterns, but it's a ubiquitous and fundamental feature of how memories are stored in the cortex. At any time, a piece can activate the whole. This is the essence of auto-associative memories.

Your neocortex is a complex biological auto-associative memory. During each waking moment, each functional region is essentially waiting vigilantly for familiar patterns or pattern fragments to come in. You can be in deep thought about something, but the instant your friend appears your thoughts switch to her. This switch isn't something you chose to do. The mere appearance of your friend forces your brain to start recalling patterns associated with her. It's unavoidable. After an interruption we frequently have to ask, "What was I thinking about?" A dinner conversation with friends follows a circuitous route of associations. The talk may start with the food in front of you, but the salad evokes an associated memory of your mother's salad at your wedding, which leads to a memory of someone else's wedding, which leads to a memory of where they went on their honeymoon, to the political problems in that part of the world, and so on. Thoughts and memories are associatively linked, and again, random thoughts never really occur. Inputs to the brain auto-associatively link to themselves, filling in the present, and auto-associatively link to what normally follows next. We call this chain of memories *thought,* and although its path is not deterministic, we are not fully in control of it either.

Now we can consider the third major attribute of neocortical memory: how it forms what are called invariant representations. I will cover the basic ideas of invariant representations in this chapter and, in chapter 6, the details of how the cortex creates them.

A computer's memory is designed to store information exactly as it is presented. If you copy a program from a CD to a hard disk, every byte is copied with 100 percent fidelity. A single error or discrepancy between the two copies might cause the program to crash. The memory in our neocortex is different. Our brain does not remember exactly what it sees, hears, or feels. We don't remember or recall things with complete fidelity—not because the cortex and its neurons are sloppy or error-prone but because the brain remembers the important relationships in the world, independent of the details. Let's look at several examples to illustrate this point.

As we saw in chapter 2, simple auto-associative memory models have been around for decades and, as I described above, the brain recalls memories auto-associatively. But there is a big difference between the auto-associative memories built by neural network researchers and those in the cortex. Artificial auto-associative memories do not use invariant representations and therefore they fail in some very basic ways. Imagine I have a picture of a face formed by a large collection of black-and-white dots. This picture is a pattern, and if I have an artificial auto-associative memory I can store many pictures of faces in the memory. Our artificial auto-associative memory is robust in that if I give it half a face or just a pair of eyes, it will recognize that part of the image and fill in the missing parts correctly. This exact experiment has been done several times. However, if I move each dot in the picture five pixels to the left, the memory completely fails to recognize the face. To the artificial auto-associative memory, it is a completely novel pattern, because none of the pixels between the previously stored pattern and the new pattern are aligned. You and I, of course, would have no difficulty seeing the shifted pattern as the same face. We probably wouldn't even notice the change. Artificial auto-associative memories fail to recognize patterns if they are moved, rotated, rescaled, or transformed in any of a thousand other ways, whereas our brains handle these variations with ease. How can we perceive something as being the same or constant when the input patterns representing it are novel and changing? Let's look at another example.

You are probably holding a book in your hands right now. As you move the book, or change the lighting, or reposition yourself in your chair, or fixate your eyes on different parts of the page, the pattern of light falling on your retina changes completely. The visual input you receive is different moment by moment and never repeats. In fact, you could hold this book for a hundred years and not once would the pattern on your retina, and therefore the pattern entering your brain, be exactly the same. Yet not for an instant do you have any doubt that you are holding a book, indeed the same book. Your brain's internal pattern representing "this book" does not change even though the stimuli informing you it's there are in constant flux. Hence we use the term *invariant representation* to refer to the brain's internal representation.

For another example, think of a friend's face. You recognize her every time you see her. It happens automatically in less than a second. It doesn't matter if she is two feet away, three feet away, or across the room. When she is close, her image occupies most of your retina. When she is far away, her image occupies a small portion of your retina. She can be facing you, turned a little to the side, or in profile. She might be smiling, squinting, or yawning. You might see her in bright light, in shade, or under strangely angled disco lights. Her visage can appear in countless positions and variations. For each one, the pattern of light falling on your retina is unique, yet in every case you know instantly that you are looking at her.

Let's pop the hood and look at what's going on in your brain to perform this amazing feat. We know from experiments that if we monitor the activity of neurons in the visual input area of your cortex, called V1, the pattern of activity is different for each different view of her face. Every time the face moves or your eyes make a new fixation, the pattern of activity in V1 changes, much like the changing pattern on the retina. However, if we monitor the activity of cells in your face recognition area—a functional region that's several steps higher than V1 in the cortical hierarchy—we find stability. That is, some set of the cells in the face recognition area remain active as long as your friend's face is anywhere in your field of vision (or even being conjured in your mind's eye), regardless of its size, position, orientation, scale, and expression. This stability of cell firing is an invariant representation.

Introspectively, this task seems so easy as to be hardly worth calling it a problem. It's as automatic as breathing. It seems trivial because we aren't consciously aware it is happening. And in some sense, it *is* trivial because our brains can solve it so quickly (remember the one hundred–step rule). However, the problem of understanding how your cortex forms invariant representations remains one of the biggest mysteries in all of science. How difficult, you ask? So much so that no one, not even using the most powerful computers in the world, has been able to solve it. And it isn't for a lack of trying.

Speculation on this problem has an ancient pedigree. It traces back to Plato, twenty-three centuries ago. Plato wondered how people are able to think and know about the world. He pointed out that real-world instances of things and ideas are always imperfect and are always different. For example, you have a concept of a perfect circle, yet you have never actually seen one. All drawings of circles are imperfect. Even if drafted with a geometer's compass a so-called circle is represented by a dark line, whereas the circumference of a true circle has no thickness at all. How then did you ever acquire the concept of a perfect circle? Or to take a more worldly case, think about your concept of dogs. Every dog you've ever seen is different from every other, and every time you see the same individual dog you see a different view of it. All dogs are different and you can never see any particular dog exactly the same way twice. Yet all of your various experiences with dogs get funneled into a mental concept of "dog" that is stable across all of them. Plato was perplexed. How is it possible that we learn and apply concepts in this world of infinitely various forms and ever-shifting sensations?

Plato's solution was his famous Theory of Forms. He concluded that our higher minds must be tethered to some transcendent plane of superreality, where fixed, stable ideas (Forms with a capital F) exist in timeless perfection. Our souls come from this mystical place before birth, he decided, which is where they learned about the Forms in the first place. After we're born we retain latent knowledge of them. Learning and understanding happen because real-world forms remind us of the Forms to which they correspond. You are able to know about circles and dogs because they respectively trigger your soul memories of Circle and Dog.

It's all quite loopy from a modern perspective. But if you strip away the high-flown metaphysics, you can see that he was really talking about invariance. His system of explanation was wildly off the mark, but his intuition that this was one of the most important questions we can ask about our own nature was a bull's-eye.

Lest you get the impression that invariance is all about vision, let's look at some examples in other senses. Consider your tactile sense. When you reach into your car's glove compartment to find your sunglasses, your fingers only have to brush against them for you to know you've found them. It doesn't matter which part of your hand makes the contact; it can be your thumb, any part of any finger, or your palm. And the contact can be with any part of the glasses, whether it's a lens, temple, hinge, or part of the frame. Just a second of moving any part of your hand over any portion of the glasses is sufficient for your brain to identify them. In each case, the stream of spatial and temporal patterns coming from your touch receptors is entirely different—different areas of your skin, different parts of the object—yet you snap up your sunglasses without a thought.

Or consider the sensorimotor task of putting the key in your car's ignition switch. The position of your seat, body, arm, and hand are slightly different each time. To you it feels like the same simple repetitive action day in, day out, but that's because you have an invariant representation of it in your brain. If you tried to make a robot that could enter the car and put in the key, you would quickly see how nearly impossible it is unless you made sure the robot was in the exact same position, and held the key in exactly the same way every time. And even if you could manage to do this, the robot would need to be reprogrammed for different cars. Robots and computer programs, like artificial auto-associative memories, are terrible at handling variation.

Another interesting example is your signature. Somewhere in your motor cortex, in your frontal lobe, you have an invariant representation of your autograph. Every time you sign your name, you use the same sequence of strokes, angles, and rhythms. This is true whether you sign it minutely with a fine-tipped pen, flamboyantly like John Hancock, in the air with your elbow, or clumsily with a pencil held between your toes. It comes out looking somewhat different each time, of course, especially under some of the awkward conditions I just named. Nevertheless, regardless of scale, writing implement, or combination of body parts, you always run the same abstract "motor program" to produce it.

From the signature example you can see that invariant representation in motor cortex is, in some ways, the mirror image of invariant representation in sensory cortex. On the sensory side, a wide variety of input patterns can activate a stable cell assembly that represents some abstract pattern (your friend's face, your sunglasses). On the motor side, a stable cell assembly representing some abstract motor command (catching a ball, signing your name) is able to express itself using a wide variety of muscle groups and respecting a wide variety of other constraints. This symmetry between perception and action is what we should expect if, as Mountcastle proposed, the cortex runs a single basic algorithm in all areas.

For a final example, let's return to sensory cortex and look at music again. (I like using memory of music as an example because it is easy to see all the issues the neocortex must solve.) Invariant representation in music is illustrated by your ability to recognize a melody in any key. The key a tune is played in refers to the musical scale the melody is built on. The same melody played in different keys starts on different notes. Once you choose the key for a rendition, you've determined the rest of the notes in the tune. Any melody can be played in any key. This means that each rendition of the "same" melody in a new key is actually an entirely different sequence of notes! Each rendition stimulates an entirely different set of locations on your cochlea, causing an entirely different set of spatial-temporal patterns to stream up into your auditory cortex . . . and yet you perceive the same melody in each case. Unless you have perfect pitch you cannot even distinguish the same song played in two different keys without hearing them back to back.

Think of the song "Somewhere over the Rainbow." You probably first learned it by hearing Judy Garland sing it in the movie *The Wizard of Oz,* but unless you have perfect pitch you probably can't recall the key she sang it in (A flat). If I sit down at a piano and start to play the song in a key in which you've never heard it—say, in D—it will sound like the same song. You won't notice that all the notes are different from those in the version you're familiar with. This means that your memory of the song must be in a form that ignores pitch. The memory must store the important relationships in the song, not the actual notes. In this case, the important relationships are the relative pitch of the notes, or "intervals." "Somewhere over the Rainbow" begins with an octave up, followed by a halftone down, followed by a major third down, and so on. The interval structure of the melody is the same for any rendition in any key. Your ability to easily recognize the song in any key indicates that your brain has stored it in this pitch-invariant form.

Similarly, the memory of your friend's face must also be stored in a form that is independent of any particular view. What makes her face recognizable are its relative dimensions, relative colors, and relative proportions, not how it appeared one instant last Tuesday at lunch. There are "spatial intervals" between the features of her face just as there are "pitch intervals" between the notes of a song. Her face is wide relative to her eyes. Her nose is short relative to the width of her eyes. The color of her hair and the color of her eyes have a similar relative relationship that stays constant even though in different lighting conditions their absolute colors change significantly. When you memorized her face, you memorized these relative attributes.

I believe a similar abstraction of form is occurring throughout the cortex, in every region. This is a general property of the neocortex. Memories are stored in a form that captures the essence of relationships, not the details of the moment. When you see, feel, or hear something, the cortex takes the detailed, highly specific input and converts it to an invariant form. It is the invariant form that is stored in memory, and it is the invariant form of each new input pattern that it gets compared to. Memory storage, memory recall, and memory recognition occur at the level of invariant forms. There is no equivalent concept in computers.

This brings up an interesting problem. In the next chapter I argue that an important function of the neocortex is to use its memory to make predictions. But given that the cortex stores invariant forms, how can it make specific predictions? Here are some examples to illustrate the problem and the solution.

Imagine it is 1890, and you are in a frontier town in the American West. Your sweetheart is taking the train from the East to join you in your new frontier home. You of course want to meet her at the station when she arrives. For a few weeks prior to her arrival day you keep track of when the trains come and go. There is no set schedule and as far as you can tell the train never arrives or leaves at the same time during the day. It is beginning to look as if you won't be able to predict when her train will arrive. But then you notice there is some structure to the trains' comings and goings. The train coming from the East arrives four hours after one leaves heading east. This four-hour gap is consistent day to day although the specific times vary greatly. On the day of her arrival, you keep an eye out for the eastbound train, and when you see it, you set your clock. After four hours you head for the station and meet her train just as it arrives. This parable illustrates both the problem the neocortex faces and the solution it uses to solve it.

The world as seen by your senses is never the same; like the arrival and departure time of the train, it is always different. The way you understand the world is by finding invariant structure in the constantly changing stream of input. However, this invariant structure alone is not sufficient to use as a basis for making specific predictions. Just knowing that the train arrives four hours after it departs doesn't allow you to show up on the platform exactly in time to greet your sweetheart. To make a specific prediction, the brain must combine knowledge of the invariant structure with the most recent details. Predicting the arrival time of the train requires recognizing the four-hour structure in the train schedule, and combining it with the detailed knowledge of what time the last eastbound train left.

When listening to a familiar song played on a piano, your cortex predicts the next note before it is played. But the memory of the song, as we've seen, is in a pitch-invariant form. Your memory tells you what interval is next, but says nothing, in and of itself, about the actual note. To predict the exact next note requires combining the next interval with the last specific note. If the next interval is a major third and the last note you heard was C, then you can predict the specific next note, E. You hear in your mind E, not "major third." And unless you've misidentified the song or the pianist slips up, your prediction is correct.

When you see your friend's face, your cortex fills in and predicts the myriad details of her unique image at that instant. It checks that her eyes are just right, that her nose, lips, and hair are exactly as they should be. Your cortex makes these predictions with great specificity. It can predict low-level details about her face even though you have never seen her in this particular orientation or environment before. If you know exactly where your friend's eyes and nose are, and you know the structure of her face, then you can predict exactly where her lips should be. If you know her skin is being tinged orange by the light of sunset, then you know what color her hair should appear. Once again, your brain does this by combining a memory of the invariant structure of her face with the particulars of your immediate experience.

The train schedule example is just an analogy of what is going on in your cortex, but the melody and face examples are not. The combining of invariant representations and current input to make detailed predictions is exactly what is happening. It is a ubiquitous process that happens in every region of cortex. It is how you make specific predictions about the room you are sitting in right now. It is how you are able to predict not only the words others will say, but also in what tone of voice they will say them, the accent they will use, and where in the room you expect to hear the voice come from. It is how you know precisely when your foot will hit the floor, and what it will feel like when you climb a set of stairs. It is how you can sign your name with your foot, or catch a thrown ball.

The three properties of cortical memory discussed in this chapter (storing sequences, auto-associative recall, and invariant representations) are necessary ingredients to predict the future based on memories of the past. In the next chapter I propose that making predictions is the essence of intelligence.

5

A NEW FRAMEWORK OF INTELLIGENCE

One day in April 1986 I was contemplating what it means to "understand" something. For months I had been struggling with the fundamental question What do brains do if they aren't generating behavior? What does a brain do when it is passively listening to speech? What is your brain doing right now while it is reading? Information goes into the brain but doesn't come out. What happens to it? Your behaviors at the moment are probably basic—such as breathing and eye movements—yet, as you are aware, your brain is doing a lot more than that as you read and understand these words. Understanding must be the result of neural activity. But what? What are the neurons doing when they understand?

As I looked around my office that day, I saw familiar chairs, posters, windows, plants, pencils, and so on. There were hundreds of items and features all around me. My eyes saw them as I glanced around, yet just seeing them didn't cause me to perform any action. No behavior was invoked or required, yet somehow I "understood" the room and its contents. I was doing what Searle's Chinese Room couldn't do, and I didn't have to pass anything back through a slot. I understood, but had no action to prove it. What did it mean to "understand"?

It was while pondering this dilemma that I had an "aha" insight, one of those emotionally powerful moments when suddenly what was a tangle of confusion becomes clear and understood. All I did was ask what would happen if a new object, one I had never seen before, appeared in the room—say, a blue coffee cup.

The answer seemed simple. I would notice the new object as not belonging. It would catch my attention as being new. I needn't consciously ask myself if the coffee cup was new. It would just jump out as not belonging. Underlying that seemingly trivial answer is a powerful concept. To notice that something is different, some neurons in my brain that weren't active before would have to become active. How would these neurons know that the blue coffee cup was new and the hundreds of other objects in the room were not? The answer to this question still surprises me. Our brains use stored memories to constantly make predictions about everything we see, feel, and hear. When I look around the room, my brain is using memories to form predictions about what it expects to experience before I experience it. The vast majority of predictions occur outside of awareness. It's as if different parts of my brain were saying, "Is the computer in the middle of the desk? Yes. Is it black? Yes. Is the lamp in the right-hand corner of the desk? Yes. Is the dictionary where I left it? Yes. Is the window rectangular and the walls vertical? Yes. Is sunlight coming from the correct direction for the time of day? Yes." But when some visual pattern comes in that I had not memorized in that context, a prediction is violated. And my attention is drawn to the error.

Of course, the brain doesn't talk to itself while making predictions, and it doesn't make predictions in a serial fashion. It also doesn't just make predictions about distinct objects like coffee cups. Your brain constantly makes predictions about the very fabric of the world we live in, and it does so in a parallel fashion. It will just as readily detect an odd texture, a misshapen nose, or an unusual motion. It isn't immediately apparent how pervasive these mostly unconscious predictions are, which is perhaps why we missed their importance for so long. They happen so automatically, so easily, we fail to fathom what is happening inside our skulls. I hope to impress on you the power of this idea. Prediction is so pervasive that what we "perceive"—that is, how the world appears to us—does not come solely from our senses. What we perceive is a combination of what we sense and of our brains' memory-derived predictions.

Minutes later I conceived a thought experiment to help convey what I understood at that moment. I call it the altered door experiment. Here is how it goes.

When you come home each day, you usually take a few seconds to go through your front door or whichever door you use. You reach out, turn the knob, walk in, and shut it behind you. It's a firmly established habit, something you do all the time and pay little attention to. Suppose while you are out, I sneak over to your home and change something about your door. It could be almost anything. I could move the knob over by an inch, change a round knob into a thumb latch, or turn it from brass to chrome. I could change the door's weight, substituting solid oak for a hollow door, or vice versa. I could make the hinges squeaky and stiff, or make them glide frictionlessly. I could widen or narrow the door and its frame. I could change its color, add a knocker where the peephole used to be, or add a window. I can imagine a thousand changes that could be made to your door, unbeknownst to you. When you come home that day and attempt to open the door, you will quickly detect that something is wrong. It might take you a few seconds' reflection to realize exactly what is wrong, but you will notice the change very quickly. As your hand reaches for the moved knob, you will realize that it is not in the correct location. Or when you see the door's new window, something will appear odd. Or if the door's weight has been changed, you will push with the wrong amount of force and be surprised. The point is that you will notice any of a thousand changes in a very short period of time.

How do you do that? How do you notice these changes? The AI or computer engineer's approach to this problem would be to create a list of all the door's properties and put them in a database, with fields for every attribute a door can have and specific entries for your particular door. When you approach the door, the computer would query the entire database, looking at width, color, size, knob position, weight, sound, and so on. While this may sound superficially similar to how I described my brain checking each of its myriad predictions as I glanced around my office, the difference is real and far-reaching. The AI strategy is implausible. First, it is impossible to specify in advance every attribute a door can have. The list is potentially endless. Second, we would need to have similar lists for every object we encounter every second of our lives. Third, nothing we know about brains and neurons suggests that this is how they work. And finally, neurons are just too slow to implement computer-style databases. It would take you twenty minutes instead of two seconds to notice the change as you go through the door.

There is only one way to interpret your reaction to the altered door: your brain makes low-level sensory predictions about what it expects to see, hear, and feel at every given moment, and it does so in parallel. All regions of your neocortex are simultaneously trying to predict what their next experience will be. Visual areas make predictions about edges, shapes, objects, locations, and motions. Auditory areas make predictions about tones, direction to source, and patterns of sound.

Somatosensory areas make predictions about touch, texture, contour, and temperature.

"Prediction" means that the neurons involved in sensing your door become active in advance of them actually receiving sensory input. When the sensory input does arrive, it is compared with what was expected. As you approach the door, your cortex is forming a slew of predictions based on past experience. As you reach out, it predicts what you will feel on your fingers, when you will feel the door, and at what angle your joints will be when you actually touch the door. As you start to push the door open, your cortex predicts how much resistance the door will offer and how it will sound. When your predictions are all met, you'll walk through the door without consciously knowing these predictions were verified. But if your expectations about the door are violated, the error will cause you to take notice. Correct predictions result in understanding. The door is normal. Incorrect predictions result in confusion and prompt you to pay attention. The door latch is not where it's supposed to be. The door is too light. The door is off center. The texture of the knob is wrong. We are making continuous low-level predictions in parallel across all our senses.

But that's not all. I am arguing a much stronger proposition. Prediction is not just one of the things your brain does. It is the *primary function* of the neocortex, and the foundation of intelligence. The cortex is an organ of prediction. If we want to understand what intelligence is, what creativity is, how your brain works, and how to build intelligent machines, we must understand the nature of these predictions and how the cortex makes them. Even behavior is best understood as a by-product of prediction.

I don't know who was the first person to suggest that prediction is key to understanding intelligence. In science and industry no one invents anything completely new. Rather, people see how existing ideas fit into new frameworks. The components of a new idea are usually floating around in the milieu of scientific discourse prior to its discovery. What is usually new is the packaging of these components into a cohesive whole. Similarly, the idea that a primary function of the cortex is to make predictions is not entirely new. It has been floating around in various forms for some time. But it has not yet assumed its rightful position at the center of brain theory and the definition of intelligence.

Ironically, some of the pioneers of artificial intelligence had a notion of computers building a model of the world and using it to make predictions. In 1956, for example, D. M. Mackay argued that intelligent machines should have an "internal response mechanism" designed to "match what is received." He didn't use the words "memory" and "prediction" but he was thinking along the same lines.

Since the mid-1990s, terms such as *inference, generative models,* and *prediction* have crept into the scientific nomenclature. They all refer to related ideas. As an example, in his 2001 book, *i of the vortex,* Rodolfo Llinas, at the New York University School of Medicine, wrote, "The capacity to predict the outcome of future events—critical to successful movement—is, most likely, the ultimate and most common of all global brain functions." Scientists such as David Mumford at Brown University, Rajesh Rao at the University of Washington, Stephen Grossberg at Boston University, and many more have written and theorized about the role of feedback and prediction in various ways. There is an entire subfield of mathematics devoted to Bayesian networks. Named after Thomas Bayes, an English minister born in 1702 who was a pioneer in statistics, Bayesian networks use probability theory to make predictions.

What has been lacking is putting these disparate bits and pieces into a coherent theoretical framework. This, I argue, has not been done before, and it is the goal of this book.

Before we get into detail about how the cortex makes predictions, let's consider some additional examples. The more you think about this idea, the more you'll realize that prediction is pervasive and the basis for how you understand the world.

This morning I made pancakes. At one point in the process, I reached under the counter to open a cabinet door. I intuitively knew, without seeing, what I would feel—in this case, the cabinet doorknob—and when I would feel it. I twisted the top of the milk container with the expectation that it would turn and then come free. I turned on the griddle expecting the knob to push in a slight amount, then turn with a certain resistance. I expected to hear the gentle *fwoomp* of the gas flame about a second later. Every minute in the kitchen I made dozens or hundreds of motions, and each one involved many predictions. I know this because if any of those common motions had had a different result from the expected one, I would have noticed it.

Every time you put your foot down while you are walking, your brain predicts when your foot will stop moving and how much "give" the material you step on will have. If you have ever missed a step on a flight of stairs, you know how quickly you realize something is wrong. You lower your foot and the moment it "passes through" the anticipated stair tread you know you are in trouble. The foot doesn't feel anything, but your brain made a prediction and the prediction was not met. A computer-driven robot would blissfully fall over, not realizing that anything was amiss, while you would know as soon as your foot continues for even a fraction of an inch beyond the spot where your brain had expected it to stop.

When you listen to a familiar melody, you hear the next note in your head before it occurs. When you listen to a favorite album, you hear the beginning of each next song a couple of seconds before it starts. What's happening? Neurons in your brain that will fire when you hear that next note fire in advance of your actually hearing it, and so you "hear" the song in your head. The neurons fire in response to memory. This memory can be surprisingly long lasting. It is not uncommon to listen to an album of music for the first time in many years and still hear the next song automatically after the previous song has ended. And it creates a pleasant sensation of mild uncertainty when you listen to your favorite CD on random shuffle; you know your prediction of the next song is wrong.

When listening to people speak, you often know what they're going to say before they've finished speaking—or at least you think you know! Sometimes we don't even listen to what the speaker actually says and instead hear what we expect to hear. (This happened to me so often when I was a child that my mother twice took me to a doctor to have my hearing checked.) You experience this in part because people tend to use common phrases or expressions in much of their conversation. If I say, "How now brown . . . ," your brain will activate neurons that represent the word *cow* before I say it (though if English is not your native language, you may have no idea what I am talking about). Of course, we don't know all the time what others are going to say. Prediction is not always exact. Rather, our minds work by making probabilistic predictions concerning what is about to happen. Sometimes we know exactly what is going to happen, other times our expectations are distributed among several possibilities. If we were eating at a table in a diner and I said, "please pass me the . . . ," your brain would not be surprised if I next said, "salt," or "pepper," or "mustard." In some sense your brain predicts all these possible outcomes at once. However, if I said, "Please pass me the sidewalk," you would know something is wrong.

Returning to music, we can see probabilistic prediction here as well. If you are listening to a song you have never heard before, you can still have fairly strong expectations. In Western music I expect a regular beat, I expect a repeated rhythm, I expect phrases to last the same number of measures, and I expect songs to end on the tonic pitch. You may not know what these terms mean, but—assuming you have listened to similar music—your brain automatically predicts beats, repeated rhythms, completion of phrases, and ends of songs. If a new song violates these principles, you know immediately that something is wrong. Think about this for a second. You hear a song that you have never heard before, your brain experiences a pattern it has never experienced before, and yet you make predictions and can tell if something is wrong. The basis of these mostly unconscious predictions is the set of memories that are stored in your cortex. Your brain can't say exactly what will happen next, but it nevertheless predicts which note patterns are likely to happen and which aren't.

We have all had the experience of suddenly noticing that a source of constant background noise, such as a distant jackhammer or droning Muzak, has just ceased—yet we hadn't noticed the sound while it was ongoing. Your auditory areas were predicting its continuation, moment after moment, and as long as the noise didn't change you paid it no heed. By ceasing, it violated your prediction and attracted your attention. Here's a historical example. Right after New York City stopped running elevated trains, people called the police in the middle of the night claiming that something woke them up. They tended to call around the time the trains used to run past their apartments.

We like to say that seeing is believing. Yet we see what we expect to see as often as we see what we really see. One of the most fascinating examples of this has to do with what researchers call filling in. It may have been brought to your attention before that you have a small blind spot in each eye, where your optic nerve exits each retina through a hole called the optic disk. You have no photoreceptors in this area, so you are permanently blind in the corresponding spot in your visual field. There are two reasons why you don't usually notice this, one mundane, the other instructive. The mundane reason is that your two blind spots don't overlap, so one eye compensates for the other.

But interestingly, you still don't notice your blind spot when only one eye is open. Your visual system "fills in" the missing information. When you close one eye and look at a richly woven Turkish carpet or the wavy contours of wood grain in a cherry tabletop, you don't see a hole. Entire nodes in the carpet, whole dark knots in the wood grain are constantly winking out of your retina's view as your blind spot happens to cover them, but your experience is of a seamless stretch of textures and colors. Your visual cortex is drawing on memories of similar patterns and is making a continuous stream of predictions that fill in for any missing input.

Filling in occurs in all parts of the visual image, not just your blind spot. For example, I show you a picture of a shore with a driftwood log lying on some rocks. The boundary between the rocks and the log is clear and obvious. However, if we magnify the image, you will see that the rocks and the log are similar in texture and color where they meet. In the enlarged view, the edge of the log isn't distinguishable from the rocks at all. If we look at the entire scene, the edge of the log is clear, but in reality we inferred the edge from the rest of the image. When we look at the world, we perceive clean lines and boundaries separating objects, but the raw data entering our eyes are often noisy and ambiguous. Our cortex fills in the missing or messy sections with what it thinks should be there. We perceive an unambiguous image.

Prediction in vision is also a function of the way your eyes move. In chapter 3, I mentioned saccades. About three times every second, your eyes fixate on one point, then suddenly jump to another point. Generally you are not aware of these movements, and you don't normally consciously control them. And each time your eyes fixate on a new point, the pattern entering your brain from the eyes changes completely from the last fixation. Thus, three times a second your brain sees something completely different. Saccades are not entirely random. When you look at a face your eyes typically fixate first on one eye, then on the other, going back and forth and occasionally fixating on the nose, mouth, ears, and other features. You perceive just "face," but the eyes see eye, eye, nose, mouth, eye, and so on. I realize it doesn't feel this way to you. What you're aware of is a continuous view of the world, but the raw data entering your head are as jerky as a badly wielded Camcorder.

Now imagine that you met someone with an extra nose where an eye should be. Your eyes fixate first on the one eye and then saccade to the second eye, but instead of seeing an eye you see a nose. You would definitely know that something was wrong. For this to happen, your brain has to have an expectation or prediction of what it is about to see. When you predict eye but see nose, the prediction is violated. So several times a second, concurrent with every saccade, your brain makes a prediction about what it will see next. When that prediction is wrong, your attention is immediately aroused. This is why we have difficulty not looking at people with deformities. If you saw a person with two noses, wouldn't you have trouble not staring? Of course, if you lived with that person, then after a period of time you would get used to two noses and not notice it as unusual anymore.

Think about yourself right now. What predictions are you making? As you turn the pages of this book, you have expectations that the pages bend a certain amount and move in predictable ways that are different from the way the cover moves. If you are sitting, you are predicting that the feelings of pressure on your body will persist; but if the seat turned wet, began drifting backward, or underwent any other unexpected change, you would stop paying attention to the book and try to figure out what is happening. If you spend some time observing yourself, you can begin to understand that your perception of the world, your understanding of the world, is intimately tied to prediction. Your brain has made a model of the world and is constantly checking that model against reality. You know where you are and what you are doing by the validity of this model.

Prediction is not limited to patterns of low-level sensory information like seeing and hearing. Up to now I've limited the discussion to such examples because they are the easiest way to introduce this framework for understanding intelligence. However, according to Mountcastle's principle, what is true of low-level sensory areas must be true for all cortical areas. The human brain is more intelligent than that of other animals because it can make predictions about more abstract kinds of patterns and longer temporal pattern sequences. To predict what my wife will say when she sees me, I must know what she has said in the past, that today is Friday, that the recycling bin has to be put on the curb on Friday nights, that I didn't do it on time last week, and that her face has a certain look. When she opens her mouth, I have a pretty strong prediction of what she will say. In this case, I don't know what the exact words will be, but I do know she will be reminding me to take out the recycling. The important point is that higher intelligence is not a different kind of process from perceptual intelligence. It rests fundamentally on the same neocortical memory and prediction algorithm.

Notice that our intelligence tests are in essence prediction tests. From kindergarten through college, I.Q. tests are based on making predictions. Given a sequence of numbers, what should the next number be? Given three different views of a complex object, which of the following is also a view of the object? Word A is to word B as word C is to what word?

Science is itself an exercise in prediction. We advance our

knowledge of the world through a process of hypothesis and testing. This book is in essence a prediction about what intelligence is and how brains work. Even product design is fundamentally a predictive process. Whether designing clothes or mobile phones, designers and engineers try to predict what competitors will do, what consumers will want, how much a new design will cost, and what fashions will be in demand.

Intelligence is measured by the capacity to remember and predict patterns in the world, including language, mathematics, physical properties of objects, and social situations. Your brain receives patterns from the outside world, stores them as memories, and makes predictions by combining what it has seen before and what is happening now.

At this point you might be thinking: "I accept that my brain makes predictions and I can be intelligent just lying in the dark. As you point out, I don't need to act in order to understand or be intelligent. But aren't situations like that the exception? Are you really arguing that intelligent understanding and behavior are completely separate? In the end, isn't behavior, not prediction, what makes us intelligent? After all, behavior is the ultimate determiner of survival."

This is a fair question and of course, in the end, behavior is what matters most to the survival of an animal. Prediction and behavior are not completely separate, but their relationship is subtle. First, the neocortex appeared on the evolutionary scene after animals already evolved sophisticated behaviors. Therefore, the survival value of the cortex must first be understood in terms of the incremental improvements it could bestow upon the animals' existing behaviors. Behavior came first, then intelligence. Second, most of what we sense is heavily dependent on what we do and how we move in the world. Therefore prediction and behavior are closely related. Let's look at these issues.

Mammals evolved a large neocortex because it gave them some survival advantage, and such an advantage must ultimately be rooted in behavior. But in the beginning, the cortex served to make more efficient use of existing behaviors, not to create entirely new behaviors. To make the case clear, we need to take a look at how our brains evolved.

Simple nervous systems emerged not long after multicellular creatures started squiggling all over the Earth, hundreds of millions of years ago, but the story of real intelligence begins more recently with our reptilian forebears. The reptiles were successful in their conquest of the land. They spread over every continent and diversified into numerous species. They had keen senses and well-developed brains that endowed them with complex behavior. Their direct descendants, today's surviving reptiles, still have them. An alligator, for example, has sophisticated senses just like you and me. It has well-developed eyes, ears, nose, mouth, and skin. It carries out complex behaviors including the ability to swim, run, hide, hunt, ambush, sun, nest, and mate.

What is the difference between a human brain and a reptile brain? A lot and a little. I say a little because, to a rough approximation, everything in a reptile's brain exists in a human brain. I say a lot because a human brain has something really important that a reptile does not have: a large cortex. You sometimes hear people refer to the "old" brain or the "primitive" brain. Every human has these more ancient structures in the brain, just like a reptile. They regulate blood pressure, hunger, sex, emotions, and many aspects of movement. When you stand, balance, and walk, for example, you are relying heavily on the old brain. If you hear a frightening sound, panic, and start to run, that is mostly your old brain. You don't need more than a reptile brain to do a lot of interesting and useful things. So what does the neocortex do if it isn't strictly required to see, hear, and move?

Mammals are more intelligent than reptiles because of their

neocortex. (The word itself is derived from the Latin words for "new bark" or "new rind," because the cortex literally covers the old brain.) The neocortex first appeared tens of millions of years ago and only mammals have one. What makes humans smarter than other mammals is primarily the large area of our neocortex—which expanded dramatically only a couple of million years ago. Remember, the cortex is built using a common repeated element. The human cortical sheet is the same thickness and has very nearly the same structure as the cortex in our mammal relatives. When evolution makes something big very quickly, as it did with human cortex, it does so by copying an existing structure. We got smart by adding many more elements of a common cortical algorithm. There is a common misconception that the human brain is the pinnacle of billions of years of evolution. This may be true if we think of the entire nervous system. However, the human neocortex itself is a relatively new structure and hasn't been around long enough to undergo much long-term evolutionary refinement.

Here then is the core of my argument on how to understand the neocortex, and why memory and prediction are the keys to unlocking the mystery of intelligence. We start with the reptilian brain with no cortex. Evolution discovers that if it tacks on a memory system (the neocortex) to the sensory path of the primitive brain, the animal gains an ability to predict the future. Imagine the old reptilian brain is still doing its thing, but now sensory patterns are simultaneously fed into the neocortex. The neocortex stores this sensory information in its memory. At a future time when the animal encounters the same or a similar situation, the memory recognizes the input as similar and recalls what happened in the past. The recalled memory is compared with the sensory input stream. It both "fills in" the current input and predicts what will be seen next. By comparing the actual sensory input with recalled memory, the animal not only understands where it is but can see into the future.

Now imagine that the cortex not only remembers what the animal has seen but also remembers the behaviors the old brain performed when it was in a similar situation. We don't even have to assume the cortex knows the difference between sensations and behavior; to the cortex they are both just patterns. When our animal finds itself in the same or a similar situation, it not only sees into the future but recalls which behaviors led to that future vision. Thus, memory and prediction allow an animal to use its existing (old brain) behaviors more intelligently.

For example, imagine you're a rat learning to navigate a maze for the first time. Aroused by uncertainty or hunger, you will use the skills inherent to your old brain to explore the new environment—listening, looking, sniffing, and creeping close to the walls. All this sensory information is used by your old brain but is also passed up to your neocortex, where it is stored. At some future time, you find yourself in the same maze. Your neocortex will recognize the current input as one it has seen before and recall the stored patterns representing what happened in the past. In essence, it allows you to see a short way into the future. If you were a talking rat, you might say, "Oh, I recognize this maze, and I remember this corner." As your neocortex recalls what happened in the past, you will envision finding the cheese you saw last time you were in the maze, and how you got to it. "If I turn right here, I know what will happen next. There's a piece of cheese down at the end of this hallway. I see it in my imagination." When you scurry through the maze, you rely on older, primitive structures to carry out movements like lifting your feet and sweeping your whiskers. With your (relatively) big neocortex, you can remember the places you have been, recognize them again in the future, and make predictions about what will happen next. A lizard without a neocortex has a much poorer ability to remember the past and may have to search a maze anew every time. You (the rat) understand the world and the immediate future because of your cortical memory. You see vivid images of the rewards and dangers that lie ahead of each decision, and so you move more effectively through your world. You can literally see the future.

But notice you are not performing any particularly complex or fundamentally new behaviors. You are not building yourself a hang glider and flying to the cheese at the end of the hallway. Your neocortex is forming predictions about sensory patterns that allow you to see into the future, but your palette of available behaviors is pretty much unaffected. Your ability to scurry, clamber, and explore is still a lot like that of a lizard.

As the cortex got larger over evolutionary time, it was able to remember more and more about the world. It could form more memories, and make more predictions. The complexity of those memories and predictions also increased. But something else remarkable happened that led to the uniquely human abilities for intelligent behavior.

Human behavior transcends the old basic repertoire of moving around with ratlike skills. We have taken neocortical evolution to a new level. Only humans create written and spoken language. Only humans cook their food, sew clothes, fly planes, and build skyscrapers. Our motor and planning abilities vastly exceed those of our closest animal relatives. How can the cortex, which was designed to make sensory predictions, generate the incredibly sophisticated behavior unique to humans? And how could this superior behavior evolve so suddenly? There are two answers to this question. One is that the neocortical algorithm is so powerful and flexible that with a little bit of rewiring, unique to humans, it can create new, sophisticated behaviors. The other answer is that behavior and prediction are two sides of the same thing. Although the cortex can envision the future, it can make accurate sensory predictions only if it knows what behaviors are being performed.

In the simple example of the rat looking for the cheese, the rat remembers the maze and uses this memory to predict that it will see the cheese around the corner. But the rat could turn left or turn right; only by simultaneously remembering the cheese and the correct behavior, "turn right at the fork," can the rat make the prediction of the cheese come true. Although this is a trivial example, it gets to the essence of how sensory prediction and behavior are intimately related. All behavior changes what we see, hear, and feel. Most of what we sense at any moment is highly dependent on our own actions. Move your arm in front of your face. To predict seeing your arm, your cortex has to know that it has commanded the arm to move. If the cortex saw your arm moving without the corresponding motor command, you would be surprised. The simplest way to interpret this would be to assume your brain first moves the arm and then predicts what it will see. I believe this is wrong. Instead I believe the cortex predicts seeing the arm, and this prediction is what causes the motor commands to make the prediction come true. You think first, which causes you to act to make your thoughts come true.

Now we want to look at the changes that led to humans having a greatly expanded behavioral repertoire. Are there physical differences between a monkey's cortex and a human's cortex that can explain why only humans have language and other complex behaviors? The human brain is about three times larger than the chimpanzee's. But there is more to it than "bigger is better." A key to understanding the leap in human behavior is found in the wiring between regions of cortex and parts of the old brain. Put most simply, our brains are connected up differently.

Let's take a closer look. Everyone is familiar with the brain's left and right hemispheres. But there is another division that is less well known, and it is where we need to look for human differences. All brains, especially large ones, divide the cortex into a front half and a back half. Scientists use the words *anterior* for the front and *posterior* for the back. Separating the front and the back is a large fissure called the central sulcus. The back part of the cortex contains the sections where the eyes, ears, and touch inputs arrive. It is where sensory perception largely occurs. The front part contains regions of cortex that are involved in high-level planning and thought. It also contains the motor cortex, the section of brain most responsible for moving muscles and therefore creating behavior.

As the primate neocortex became larger over time, the anterior half got disproportionately larger, especially so in humans. Compared with other primates and early hominids, we have enormous foreheads designed to contain our very large anterior cortex. But this enlargement alone is not enough to explain the improvement in our motor ability as compared with that of other creatures. Our ability to make exceptionally complex movements stems from the fact that our motor cortex makes many more connections with the muscles in our bodies. In other mammals, the front cortex plays a less direct role in motor behavior. Most animals rely largely on the older parts of the brain for generating their behavior. In contrast, the human cortex usurped most of the motor control from the rest of the brain. If you damage the motor cortex of a rat, the rat may not have noticeable deficits. If you damage the motor cortex of a human, he or she becomes paralyzed.

People often ask me about dolphins. Don't they have huge brains? The answer is yes; a dolphin has a large neocortex. Dolphin cortex has a simpler structure (three layers versus our six) than a human neocortex, but by any other measure it is large. It is likely a dolphin can remember and understand lots of things. It can recognize other individual dolphins. It probably has an excellent memory of its own life, in an autobiographical sense. It probably knows every nook and cranny of the ocean it's ever been to. But although they exhibit some sophisticated behavior, dolphins don't come close to our own. So we can surmise their cortex has a less-dominant influence on their behavior. The point is that the cortex evolved primarily to provide a memory of the world. An animal with a large cortex could perceive the world much as you and I do. But humans are unique in the dominant, advanced role the cortex plays in our behavior. It is why we have complex language and intricate tools whereas other animals don't. It is why we can write novels, surf the Internet, send probes to Mars, and build cruise ships.

Now we can see the entire picture. Nature first created animals such as reptiles with sophisticated senses and sophisticated but relatively rigid behaviors. It then discovered that by adding a memory system and feeding the sensory stream into it, the animal could remember past experiences. When the animal found itself in the same or a similar situation, the memory would be recalled, leading to a prediction of what was likely to happen next. Thus, intelligence and understanding started as a memory system that fed predictions into the sensory stream. These predictions are the essence of understanding. To know something means that you can make predictions about it.

The cortex evolved in two directions. First it got larger and more sophisticated in the types of memories it could store; it was able to remember more things and make predictions based on more complex relationships. Second, it started interacting with the motor system of the old brain. To predict what you will hear, see, and feel next, it needed to know what actions were being taken. With humans the cortex has taken over most of our motor behavior. Instead of just making predictions based on the behavior of the old brain, the human neocortex directs behavior to satisfy its predictions.

The human cortex is particularly large and therefore has a massive memory capacity. It is constantly predicting what you will see, hear, and feel, mostly in ways you are unconscious of. These predictions are our thoughts, and, when combined with sensory input, they are our perceptions. I call this view of the brain the *memory-prediction framework* of intelligence.

If Searle's Chinese Room contained a similar memory system that could make predictions about what Chinese characters would appear next and what would happen next in the story, we could say with confidence that the room understood Chinese and understood the story. We can now see where Alan Turing went wrong. Prediction, not behavior, is the proof of intelligence.

We are now ready to delve into the details of this new idea of the memory-prediction framework of the brain. To make predictions of future events, your neocortex has to store sequences of patterns. To recall the appropriate memories, it has to retrieve patterns by their similarity to past patterns (auto-associative recall). And, finally, memories have to be stored in an invariant form so that the knowledge of past events can be applied to new situations that are similar but not identical to the past. How the physical cortex accomplishes these tasks, plus a fuller exploration of its hierarchy, is the subject of the next chapter.

6

HOW THE CORTEX WORKS

Trying to figure out how the brain works is like solving a giant jigsaw puzzle. You can approach it in one of two ways. Using the "top-down" approach, you start with the image of what the solved puzzle should look like, and use this to decide which pieces to ignore and which pieces to search for. The other approach is "bottom-up," where you focus on the individual pieces themselves. You study them for unusual features and look for close matches with other puzzle pieces. If you don't have a picture of the puzzle's solution, the bottom-up method is sometimes the only way to proceed.

The "understand-the-brain" jigsaw puzzle is particularly daunting. Lacking a good framework for understanding intelligence, scientists have been forced to stick with the bottom-up approach. But the task is Herculean, if not impossible, with a puzzle as complex as the brain. To get a sense of the difficulty, imagine a jigsaw puzzle with several thousand pieces. Many of the pieces can be interpreted multiple ways, as if each had an image on both sides but only one of them is the right one. All the pieces are poorly shaped so you can't be certain if two pieces fit together or not. Many of them will not be used in the ultimate solution, but you don't know which ones or how many. Every month new pieces arrive in the mail. Some of these new pieces replace older ones, as if the puzzle maker was saying, "I know you've been working with these old puzzle pieces for a few years, but they turned out to be wrong. Sorry. Use these new ones instead until future notice." Unfortunately, you have no idea what the end result will look like; worse, you may have some ideas, but they are wrong.

This puzzle analogy is a pretty good description of the difficulty we face in creating a new theory of the cortex and intelligence. The puzzle pieces are the biological and behavioral data that scientists have collected for well over one hundred years. Each month new papers are published, creating additional puzzle pieces. Sometimes the data from one scientist contradict the data from another. Because the data can be interpreted in different ways, there is disagreement over practically everything. Without a top-down framework, there is no consensus on what to look for, what is most important, or how to interpret the mountains of information that have accrued. Our understanding of the brain has been stuck in the bottom-up approach. What we need is a top-down framework.

The memory-prediction model can play this role. It can show us how to start putting pieces of the puzzle together. To make predictions, your cortex needs a way to memorize and store knowledge about sequences of events. To make predictions of novel events, the cortex must form invariant representations. Your brain needs to create and store a model of the world as it is, independent from how you see it under changing circumstances. Knowing what the cortex must do guides us to understanding its architecture, especially its hierarchical design and six-layered form.

As we explore this new framework, presented here for the first time, I will get into a level of detail that may be challenging for some readers. Many of the concepts you are about to encounter are unfamiliar, even to experts in neuroscience. But with a bit of effort, I believe anyone can learn the fundamentals of this new framework. Chapters 7 and 8 of this book are far less technical and explore the wider implications of the theory.

Our puzzle-solving journey can now turn to looking for biological details that support the memory-prediction hypothesis; this is like being able to set aside a large percentage of the puzzle pieces, knowing that the relatively few remaining pieces are going to reveal the ultimate solution. Once we know what to look for, the task becomes manageable.

At the same time, I want to stress that this new framework is incomplete. There are many things I don't yet understand. But there are many things I do, based on deductive reasoning, experiments carried out in many different laboratories, and known anatomy. In the last five to ten years, researchers from many subspecialties in neuroscience have been exploring ideas similar to mine, although they use different terminology and have not, as far as I know, tried to put these ideas into an overarching framework. They do talk about top-down and bottom-up processing, how patterns propagate through sensory regions of the brain, and how invariant representations might be important. For example, Gabriel Kreiman and Christof Koch, neuroscientists at Caltech, with the neurosurgeon Itzhak Fried at UCLA, have found cells that fire whenever a person sees a picture of Bill Clinton. One of my goals is to explain how those Bill Clinton cells come into being. Of course, all theories need to make predictions that can be tested in the laboratory. I've suggested a number of these predictions in the appendix. Now that we know what to look for, this very complex system won't look so complex anymore.

In the following sections of this chapter, we will probe deeper and deeper into how the memory-prediction model of the cortex works. We will start with the large-scale structure and large-scale function of the neocortex and work toward understanding the smaller pieces and how they fit into the big picture.

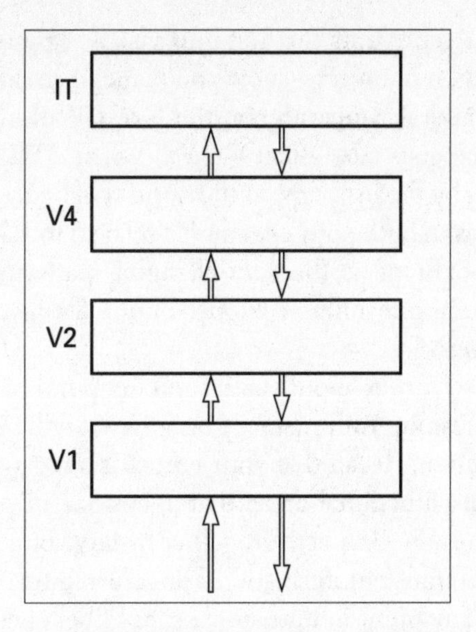

Figure 1. The first four visual regions in the recognition of objects.

INVARIANT REPRESENTATIONS

Earlier, I painted a picture of the cortex as a sheet of cells the size of a dinner napkin, as thick as six business cards, where the connections between various regions give the whole thing a hierarchical structure. Now I want to draw another picture of the cortex that highlights its hierarchical connectivity. Imagine we cut up the dinner napkin into separate functional regions—sections of cortex that specialize in certain tasks—and stack those regions on top of each other like pancakes. If you cut through this stack and view it from the side, you get figure 1. The cortex doesn't actually look like this, mind you, but the image will help you visualize how the information is flowing. I have shown four cortical regions in which sensory input enters at the bottom, the lowest region, and flows upward from region to region. Notice information flows both ways.

Figure 1 represents the first four visual regions involved in the recognition of objects—how you come to see and recognize a cat, a cathedral, your mother, the Great Wall of China, you name it. Biologists label them V1, V2, V4, and IT. Visual input represented by the up arrow at the bottom of figure 1 originates in the retinas in both your eyes and is relayed to V1. This input can be thought of as the ever-changing patterns carried on approximately one million axons, bundled together to form your optic nerve.

We talked earlier about spatial and temporal patterns, but it is worth refreshing your memory because we will be referring to them here often. Recall that your cortex is a big sheet of tissue that contains functional areas that specialize in certain tasks. These regions are connected together by large bundles of axons or fibers that transfer information from one region to another, all at once. At any point in time, some set of fibers will fire electrical pulses, again called action potentials or spikes, while others will remain quiet. The collective activity on a bundle of fibers is what is meant by *pattern.* The pattern arriving at V1 can be spatial, as when your eyes fall for an instant on an object, and temporal, as when your eyes move across the object.

As noted earlier, about three times a second your eyes make a quick movement, called a saccade, and a stop, called a fixation. If a scientist were to fit you with a device that tracks eye movements, you'd be surprised to discover how jerky your saccades are, given that you experience vision as continuous and stable. Figure 2a shows how one person's eyes moved while looking at a face. Notice how the fixations are not random. Now imagine you could see the pattern of activity arriving at V1 from this person's eyes. It changes completely with each saccade. Several times a second the visual cortex sees a completely new pattern.

You might think, "OK, but it's still the same face, just shifted." There is some truth to this, but not as much as you think. The light receptors in your retina are unevenly distributed. They are

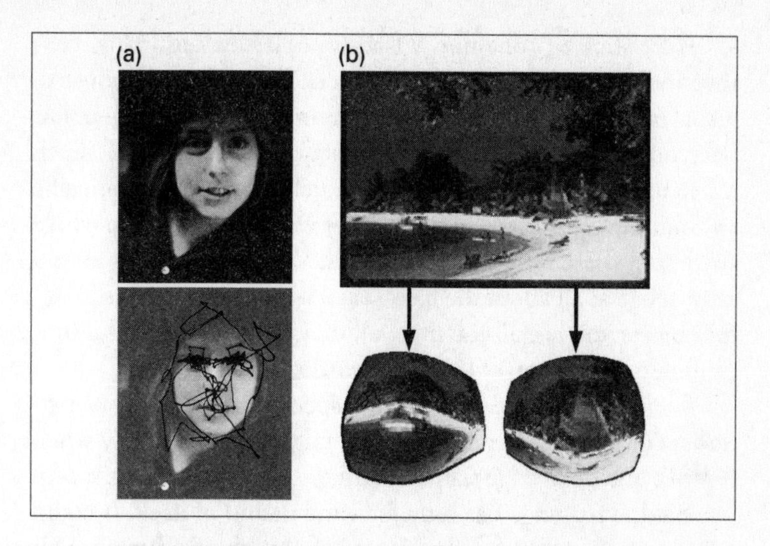

Figure 2a. How the eye makes saccades across a human face.
Figure 2b. Distortion caused by the uneven distribution of receptors in the retina.

densely concentrated in the fovea at the center, and get gradually sparser out in the periphery. In contrast, the cells in the cortex are evenly distributed. The result is that the retinal image relayed onto the primary visual area, V1, is highly distorted. When your eyes fixate on the nose of a face versus on an eye of the same face, the visual input is very different, as though it is being viewed through a distorting fisheye lens that is jerking violently to and fro. Yet when you see the face, it doesn't appear distorted, and it doesn't appear to be jumping around. Most of the time you aren't even aware that the retinal pattern has changed at all, let alone so dramatically. You just see "face." (Figure 2b shows this effect on a view of a beach landscape.) This is a restatement of the mystery of invariant representation we talked about in chapter 4, on memory. What you "perceive" is not what V1 sees. How does your brain ever know it is looking at the same face, and why don't you know the inputs are changing and distorted?

If we stick a probe into V1 and watch how individual cells respond, we find that any particular cell fires only in response to visual input from a tiny part of your retina. This experiment has been done many times and is a mainstay of vision research. Each V1 neuron has a so-called receptive field that is highly specific to a minute part of your total field of vision—that is, the whole world out there in front of your eyes. V1 cells seem to have no knowledge at all about the faces, cars, books, or other meaningful objects you see all the time; all they "know" about is a tiny, pinhole-size portion of the visual world.

Each V1 cell is also tuned for specific kinds of input patterns. For instance, a particular cell might fire vigorously when it sees a line or an edge slanted at thirty degrees within its receptive field. That edge has little meaning in and of itself. It could be part of any object—a floorboard, the trunk of a distant palm tree, the side of a letter *M,* or any of seemingly infinite other possibilities. With each new fixation, the cell's receptive field comes to rest on a new and entirely different portion of visual space. On some fixations the cell will fire strongly, on others it will fire weakly or not at all. Thus each time you make a saccade, many cells in V1 are likely to change their activity.

However, something magical happens if you stick a probe into the top region shown in figure 1, region IT. Here we find some cells that become active and stay active when entire objects appear anywhere in the visual field. For example, we might find a cell that fires robustly whenever a face is visible. This cell stays active as long as your eyes are looking at a face anywhere in your field of vision. It doesn't switch on and off with each saccade as do the cells in V1. This IT cell's receptive field covers most of the visual space and is tuned to fire when it sees faces.

Let's recast the mystery. In the course of spanning four cortical stages from retina to IT, cells have changed from being rapidly changing, spatially specific, tiny-feature recognition cells, to being constantly firing, spatially nonspecific, object recognition cells. The IT cell tells us we are seeing a face somewhere in our field of view. This cell, commonly called a face cell, will fire no matter whether the face is tilted, rotated, or partially occluded. It is part of an invariant representation for "face."

Writing these words makes it seem so simple. Four quick stages and, voilà, we recognize a face. No computer program or mathematical formula has solved this problem with anywhere near the robustness and generality of a human brain. Yet we know the brain solves it in a few steps, so the answer can't be that difficult. One of the primary goals of this chapter is to explain how a face cell, Bill Clinton's or otherwise, comes about. We will get there, but we have to cover a lot of other ground first.

Take another look at figure 1. You can see that information also flows from higher to lower regions via a network of feedback connections. These are bundles of axons that go from higher regions like IT to lower regions like V4, V2, and V1. Moreover, there are as many if not more feedback connections in visual cortex as there are feedforward connections.

For many years most scientists ignored these feedback connections. If your understanding of the brain focused on how the cortex took input, processed it, and then acted on it, you didn't need feedback. All you needed were feedforward connections leading from sensory to motor sections of the cortex. But when you begin to realize that the cortex's core function is to make predictions, then you have to put feedback into the model; the brain has to send information flowing back toward the region that first receives the inputs. Prediction requires a comparison between what is happening and what you expect to happen. What is actually happening flows up, and what you expect to happen flows down.

The same feedforward–feedback process is occurring in all your cortical areas involving all your senses. Figure 3 shows our

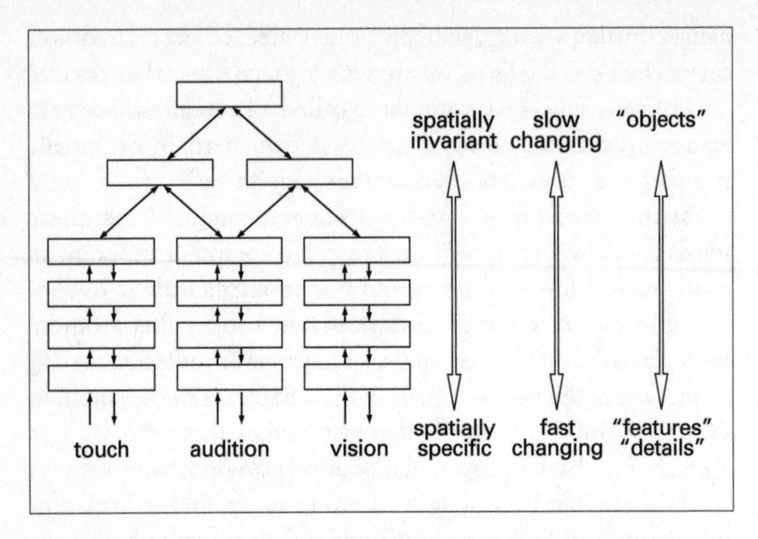

Figure 3. Forming invariant representations in hearing, vision, and touch.

visual pancake stack alongside similar stacks for hearing and touch. It also shows a few higher cortical regions, the association areas, that receive and integrate inputs from a number of different senses—such as hearing plus touch plus vision. While figure 1 is based on known connectivity between four known regions of cortex, figure 3 is a purely conceptual diagram, with no attempt to capture actual cortical regions. In a real human brain dozens of cortical regions are interconnected in all sorts of ways. In fact, the majority of the human cortex consists of association areas. The cartoon characterization shown here and in subsequent figures is meant to help you understand what is going on without misleading you in any significant way.

The transformation—from fast changing to slow changing and from spatially specific to spatially invariant—is well documented for vision. And although there is a smaller body of evidence to prove it, many neuroscientists believe you'd find the same thing happening in all the sensory areas of your cortex, not just in vision.

Take hearing. When someone speaks to you, the changes in sound pressure occur very rapidly; the patterns entering the primary auditory area, called A1, change just as rapidly. Yet if we could poke a probe higher up into your auditory stream, we would find invariant cells that respond to words or even, in some cases, phrases. Your auditory cortex might have a group of cells that fire when you hear "thank you," and another group of cells that fire for the phrase "good morning." Such cells would stay active for the whole length of an utterance, assuming that you recognize the phrase.

Patterns received by the first auditory area can vary widely. A word can be spoken with different accents, in different pitches, or at different speeds. But higher up in the cortex, those low-level features don't matter; a word is a word regardless of the acoustic details. The same thing goes for music. You can hear "Three Blind Mice" played on the piano, on the clarinet, or sung by a child, and your A1 region receives a completely different pattern in each case. But a probe stuck into a higher auditory region should find cells that fire steadily every time "Three Blind Mice" is being played, without concern for the instrument, tempo, or other details. This particular experiment has not been done, of course, since it's too invasive to perform on humans, but, if you accept that there must be a common cortical algorithm, you can be sure such cells exist. We see the same kind of feedback, prediction, and invariant recall in auditory cortex as we saw in the visual system.

Finally, touch should behave the same way. Again, the definitive experiments have not been done, although research is under way using monkeys in high-resolution brain-imaging machines. As I sit here writing, I have a pen in my hand. I touch the pen's cap and my finger caresses its metal pocket clip. The patterns entering my somatosensory cortex from the touch sensors in my skin are changing rapidly as my fingers move, yet I perceive a constant pen. At one moment I might flex the metal clip with my fingers, the next moment I do so with a different set of fingers, or even with my lips. These are very different inputs, arriving at different locations in the primary somatosensory cortex. However, our probe would once again find cells in regions several steps removed from the primary input that respond invariantly to "pen." They would stay active while I fondled the pen and wouldn't care exactly which fingers or parts of my body I used to touch it.

Think of this. With hearing and touch you can't recognize an object with a momentary input. The pattern coming from your ears or the touch sensors in your skin does not contain sufficient information at any one point in time to tell you what you are hearing or feeling. When you perceive a series of auditory patterns such as a melody, a spoken word, or a slamming door, and when you perceive a tactile object such as a pen, the only way to do so is by using the flow of input over time. You can't recognize a melody by hearing one note, and you can't recognize the feel of a pen with one touch. Therefore the neural activity corresponding to the mental perception of objects, such as spoken words, must last longer in time than the individual input patterns. This is just another way of reaching the same conclusion that the higher up in the cortex you go, the fewer changes over time you should see.

Vision is also a time-based input stream and works in the same general way as hearing and touch, but because we have the ability to recognize individual objects with a single fixation, it confuses the picture. Indeed, this ability to recognize spatial patterns during a brief fixation has for many years misled researchers who work on machine and animal vision. They have generally ignored the critical nature of time. While humans can, under laboratory conditions, be made to recognize objects without moving their eyes, this is not the norm. Normal vision, such as you reading this book, requires constant eye movement.

INTEGRATING THE SENSES

What about the association areas? So far, we have seen how information flows up and down a particular sensory area of cortex. The downward flow fills in the current input and makes predictions about what we will experience next. The same process occurs between senses—that is, between sight, sound, and touch. For example, something I hear can lead to a prediction of what I should see or feel. Right now I am writing in my bedroom. Our family cat, Keo, has a collar that jingles as she walks. I hear her jingle approaching from the hallway. From this auditory input I recognize my cat, turn my head toward the hallway, and in walks Keo. I expected to see her based on her sound. If Keo hadn't walked in, or another animal had appeared, I would have been surprised. In this example, an auditory input first created an auditory recognition of Keo. The information flowed up the auditory hierarchy to an association area that connects vision with hearing. The representation then flowed back down the auditory and visual hierarchies, leading to both auditory and visual predictions. Figure 4 illustrates this.

This kind of multisensory prediction is occurring all the time. I bend out the clip on my pen, I feel the clip slip from my fingers, and I expect to hear a snapping sound as the clip hits the barrel of the pen. If I didn't hear the snap following the release of the clip, I would be surprised. My brain predicts precisely when I will hear the sound and what it will be like. For this prediction to happen, information flowed up the somatosensory cortex and flowed back down both the somatosensory and auditory cortex leading to a prediction of hearing and feeling a snap.

Another example: I ride my bicycle to work several days a week. On those mornings I go into my garage, pick up my bicycle, turn it around, and roll it out into the driveway. In the process of doing so, I receive many visual, tactile, and auditory

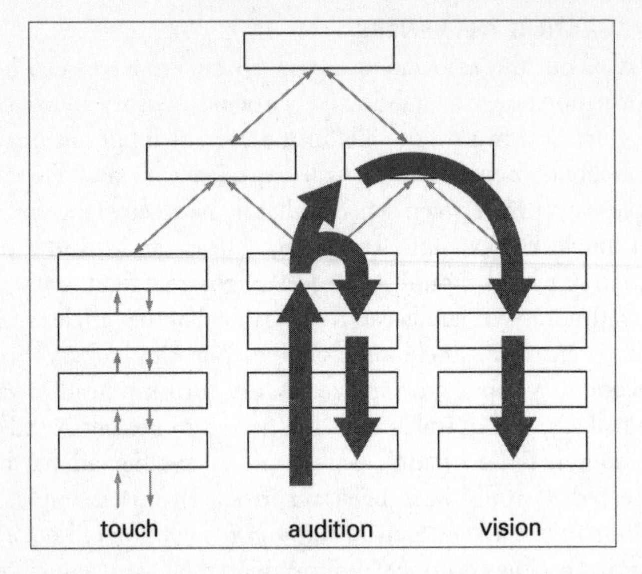

Figure 4. Information flows up and down sensory hierarchies to form predictions and create a unified sensory experience.

inputs. The bicycle hits the doorjamb, the chain rattles, a pedal hits my leg, and the wheel spins as it rubs the floor. In the process of carrying my bicycle out of the garage, my brain encounters a barrage of sight, sound, and touch sensations. Each sensory input stream makes predictions for the others in an amazingly coordinated way. Things I see lead to precise predictions about things I will feel and hear, and the other way around. Seeing the bicycle hit the doorjamb makes me expect to hear a particular sound and feel the bicycle bounce upward. Feeling the pedal hit my leg prompts me to look down and predict to see the pedal right where I felt it. The predictions are so precise that I would notice if any of these inputs were even slightly uncoordinated or unusual. The information simultaneously flows up and down the sensory hierarchies to create a unified sensory experience involving prediction in all senses.

Try this experiment. Get up from reading and do something, any activity that involves moving your body and manipulating an object. For example, go to a sink and turn on the faucet. Now, as you do this, try to notice every sound, touch sensation, and changing visual input. You will have to concentrate. Each action is intimately tied to sights, sounds, and touch sensations. Lift or turn the faucet lever and your brain expects to feel pressure on your skin and resistance in your muscles. You expect to see and feel the lever move, and you expect to see and hear the water. As the water hits the sink, you expect to hear a different sound and to see and feel the splash.

Every footstep makes a sound, which you always anticipate, whether consciously or not. Even the simple act of holding this book leads to many sensory predictions. Imagine if you felt and heard the book close but visually it stayed open. You would be shocked and confused. As we saw with the altered door thought experiment in chapter 5, you make constant predictions about the world that are coordinated across all your senses. When I concentrate on all the little sensations, I am amazed at how fully integrated our perceptual predictions are. Although these predictions may seem simple or trivial, bear in mind how pervasive they are and how they can only come about by massive coordination of patterns streaming up and down the cortical hierarchy.

Once you understand how interconnected the senses are, you are drawn to conclude that the entire neocortex, all the sensory and association areas, acts as one. Yes, we have a visual cortex, but it is just one component of a single, overarching sensory system—sights, sounds, touch, and more combined, all flowing up and down a single multibranched hierarchy.

A further point: All predictions are learned by experience. We expect pen clips to make snapping sounds in the present and future because they have done so in the past. Bicycles banging in garages look, feel, and sound to us in predictable ways. You were not born with any of this knowledge; you learned it thanks to the incredibly large capacity of your cortex for remembering patterns. If there are consistent patterns among the inputs flowing into your brain, your cortex will use them to predict future events.

Although figures 3 and 4 do not depict the motor cortex, you can imagine it as another hierarchical stack of pancakes, just like a sensory stack, connected to the sensory systems via association areas (although with more intimate connections to the somatosensory cortex for carrying out body movements). In this fashion, the motor cortex behaves in almost the same way as a sensory region. An input in any sensory area can flow up to an association area, which can lead to a pattern flowing down the motor cortex, resulting in behavior. Just as a visual input can lead to patterns flowing down the auditory and touch parts of the cortex, it can also lead to a pattern flowing down the motor section of cortex. In the former case, we interpret these downward-flowing patterns as predictions. In the motor cortex we interpret them as motor commands. As Mountcastle pointed out, the motor cortex looks like the sensory cortex. Therefore, the way the cortex processes downward-flowing sensory predictions is similar to how it processes downward-flowing motor commands.

We will see shortly that there are no pure sensory or pure motor areas in the cortex. Sensory patterns simultaneously flow in anywhere and everywhere—and then flow back down any area of the hierarchy, leading to predictions or motor behavior. Although the motor cortex has some special attributes, it is correct to think of it as just part of one large hierarchical memory-prediction system. It's almost like another sense. Seeing, hearing, touching, and acting are profoundly intertwined.

A NEW VIEW OF V1

The next step in untangling the architecture of the cortex requires looking at the cortical regions in a new way. We know the higher regions of cortical hierarchy form invariant representations. But why should this important function only occur at the top? With Mountcastle's notion of symmetry in the back of my mind, I started to explore the different ways cortical regions might be connected.

Figure 1 portrays the four classic regions of the visual pathway, V1, V2, V4, and IT, with V1 at the bottom of the stack overlain by V2, V4, and, at the top, IT. Conventionally, each is considered and shown as a single, continuous region. Thus all V1 cells supposedly do similar things, although with different parts of the visual field. All of V2's cells are doing the same kind of task. All V4 cells are similarly specialized.

In this traditional view, when the image of a face enters the V1 region, the cells therein create a rough sketch of the face in terms of simple line segments and other elementary features. The sketch is handed up to V2. Then V2 does its own thing with the image, making a slightly more sophisticated analysis of the facial features, and passes that up to V4, and so on. Invariance, and the recognition of the object, is only achieved when the input reaches the top, IT.

Unfortunately, there are some problems with this view of early cortical regions like V1, V2, and V4. Again, why should invariant representations only show up in IT? If all cortical regions perform the same function, why should IT be special?

Second, a face can appear on the left side of your V1 or on the right side of your V1 and you would recognize it. But experiments clearly show that nonadjacent patches of V1 are not directly connected; the left side of V1 can't know directly what the right side is seeing. Step back and think about this. Different parts of V1 are obviously doing something similar since they all can participate in recognizing a face, but at the same time they are physically independent. Subregions or clusters of V1 are physically disconnected but do the same thing.

Finally, experiments show that all higher regions of cortex receive converging inputs from two or more sensory regions below themselves (figure 3). In real brains, a dozen regions can converge on an association area. But in traditional renderings, lower sensory regions like V1, V2, and V4 appear to have a different kind of connectivity. Each looks as though it has but one source of input—only one arrow flowing up from the bottom—with no obvious convergence of inputs from different regions. V2 got its input from V1 and that was it. Why should some cortical regions receive converging input and others not? This, too, is inconsistent with Mountcastle's idea of a common cortical algorithm.

For these and other reasons I have come to believe that V1, V2, and V4 should not be viewed as single cortical regions. Rather, each is a collection of many smaller subregions. Let's go back to the dinner napkin analogy—a flattened version of the entire cortex. Let's say we were to use a pen to mark all the functional regions of the cortex on our cortical napkin. The largest region by far is V1, the primary visual area. Next would be V2. They are huge compared to most regions. What I am suggesting is that V1 should actually be considered as many very small regions. Instead of one big area of the napkin, we would draw many small areas that together would occupy the area normally assigned to V1. In other words, V1 is made up of numerous separate little cortical areas that are only connected to their neighbors indirectly, through regions higher up in the hierarchy. V1 would have the largest number of small subregions of any visual area. V2 would also be composed of fewer, slightly larger subregions. The same would be true of V4. But by the time you get to the top region, IT, you really would have a single region, which is why IT cells have a bird's-eye view of the entire visual world.

There is a pleasing symmetry here. Take a look at figure 5,

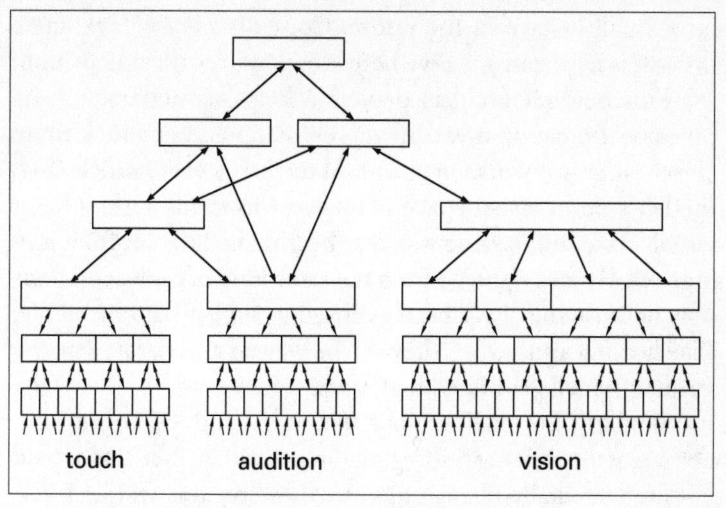

Figure 5. Alternate view of the cortical hierarchy.

which shows the same hierarchy as figure 3, except it shows the sensory hierarchies as I just described. Note that now the cortex looks similar everywhere. Pick any region and you will find many lower regions providing converging sensory input. The receiving region sends projections back to its input regions telling them what patterns they should expect to see next. Higher association areas unite information from multiple senses such as vision and touch. A lower region like a subregion of V2 unites the information from separate subregions within V1. A region doesn't know—indeed it can't know—what any of those inputs mean. A V2 subregion doesn't need to know that it is handling visual input from multiple parts of V1. An association area doesn't need to know that it is handling input from vision and hearing. Rather, the job of any cortical region is to find out how its inputs are related, to memorize the sequence of correlations between them, and to use this memory to predict how the inputs will behave in the future. Cortex is cortex. The same process is happening everywhere: a common cortical algorithm.

This new hierarchical depiction helps us understand the process of creating invariant representations. Let's look more closely at how it works in vision. At the first level of processing, the left side of visual space is different from the right side of visual space in the same way that hearing is different from seeing. Left V1 and right V1 form the same kind of representations only because they have been exposed to similar patterns in life. Like hearing and seeing, they can be viewed as separate sensory streams, which get united higher up.

Similarly, the small regions within V2 and V4 are association areas of vision. (Subregions may overlap, but this would not fundamentally change the way these regions work.) Interpreting the visual cortex this way does not contradict or change anything we know about its anatomy. Information flows up and down all branches of the hierarchical memory tree. A pattern in the left visual field can lead to a prediction in the right visual field in exactly the same way that my cat's bell can lead to a visual prediction that she is entering my bedroom.

The most important result of this new depiction of the cortical hierarchy is that now we can say each and every region of cortex forms invariant representations. In the old way of thinking, we didn't have complete invariant representations—such as faces—until inputs reached the top layer, IT, which sees the whole visual world. Now we can say that invariant representations are ubiquitous. Invariant representations are formed in every cortical region. Invariance isn't something that only magically appears when we get to higher regions of the cortex, such as IT. Every region forms invariant representations drawn from the input areas hierarchically below it. Thus the subregions of V4, V2, and V1 create invariant representations based on what flows into them. They may only see a tiny part of the world, and the vocabulary of sensory objects they deal with is more basic, but they are performing the same job as IT. Also, association regions above IT form invariant representations of patterns from multiple senses. Thus all regions of cortex form invariant representations of the world underneath them in the hierarchy. There is beauty in this.

Our puzzle has shifted. We no longer have to ask how invariant representations are formed in four steps from bottom to top. Rather we have to ask how invariant representations are formed in every single cortical region. This makes perfect sense if we take the existence of a common cortical algorithm seriously. If one region stores sequences of patterns, every region stores sequences. If one region creates invariant representations, all regions create invariant representations. Redrawing the cortical hierarchy along the lines shown in figure 5 makes this interpretation possible.

A MODEL OF THE WORLD

Why is the neocortex built as a hierarchy?

You can think about the world, move around in the world, and make predictions of the future because your cortex has built a model of the world. One of the most important concepts in this book is that the cortex's hierarchical structure stores a model of the hierarchical structure of the real world. The real world's nested structure is mirrored by the nested structure of your cortex.

What do I mean by a nested or hierarchical structure? Think about music. Notes are combined to form intervals. Intervals are combined to form melodic phrases. Phrases are combined to form melodies or songs. Songs are combined into albums. Think about written language. Letters are combined to form syllables. Syllables are combined to form words. Words are combined to form clauses and sentences. Looking at it the other way around, think about your neighborhood. It probably contains roads, schools, and houses. Houses have rooms. Each room has walls, a ceiling, a floor, a door, and one or more windows. Each of these is composed of smaller objects. Windows are made of glass, frames, latches, and screens. Latches are made from smaller parts like screws.

Take a moment to look up at your surroundings. Patterns from the retina entering your primary visual cortex are being combined to form line segments. Line segments combine to form more complex shapes. These complex shapes are combining to form objects like noses. Noses are combining with eyes and mouths to form faces. And faces are combining with other body parts to form the person who is sitting in the room across from you.

All objects in your world are composed of subobjects that occur consistently together; that is the very definition of an object. When we assign a name to something, we do so because a set of features consistently travels together. A face is a face precisely because two eyes, a nose, and a mouth always appear together. An eye is an eye precisely because a pupil, an iris, an eyelid, and so on, always appear together. The same can be said for chairs, cars, trees, parks, and countries. And, finally, a song is a song because a series of intervals always appear together in sequence.

In this way the world is like a song. Every object in the world is composed of a collection of smaller objects, and most objects are part of larger objects. This is what I mean by nested structure. Once you are aware of it, you can see nested structures everywhere. In an exactly analogous way, your memories of things and the way your brain represents them are stored in the hierarchical structure of the cortex. Your memory of your home does not exist in one region of cortex. It is stored over a hierarchy of cortical regions that reflect the hierarchical structure of the home. Large-scale relationships are stored at the top of the hierarchy and small-scale relationships are stored toward the bottom.

The design of the cortex and the method by which it learns naturally discover the hierarchical relationships in the world.

You are not born with knowledge of language, houses, or music. The cortex has a clever learning algorithm that naturally finds whatever hierarchical structure exists and captures it. When structure is absent, we are thrown into confusion, even chaos.

You can only experience a subset of the world at any moment in time. You can only be in one room of your home, looking in one direction. Because of the hierarchy of the cortex, you are able to know that you are at home, in your living room, looking at a window, even though at that moment your eyes happen to be fixated on a window latch. Higher regions of cortex are maintaining a representation of your home, while lower regions are representing rooms, and still lower regions are looking at a window. Similarly, the hierarchy allows you to know you are listening to both a song and an album of music, even though at any point in time you are hearing only one note, which on its own tells you next to nothing. It allows you to know you are with your best friend, even though your eyes are momentarily fixated on her hand. Higher regions of your cortex are keeping track of the big picture while lower areas are actively dealing with the fast-changing, small details.

Since we can only touch, hear, and see a very small part of the world at any moment in time, information flowing into the brain naturally arrives as a sequence of patterns. The cortex wants to learn those sequences that occur over and over again. In some cases, such as melodies, a sequence of patterns comes in a rigid order, the order of the intervals. Most of us are familiar with that kind of sequence. But I am going to use the word *sequence* in a more general way, closer in meaning to the mathematical term *set*. A sequence is a set of patterns that generally accompany each other but not always in a fixed order. What is important is that patterns of a sequence follow one another in time even if not in a fixed order.

Some examples should make this clear. When I look at your face, the sequence of input patterns I see is not fixed but is determined by my saccades. One time I might fixate in the order "eye eye nose mouth," and a moment later fixate in the order "mouth eye nose eye." The components of a face are a sequence. They are statistically related and tend to occur together in time, although the order may vary. If you perceive "face" while fixating on "nose," the likely next patterns would be "eye" or "mouth" but not "pen" or "car."

Each region of cortex sees a stream of such patterns. If the patterns are related in such a way that the region can learn to predict what pattern will occur next, the cortical region forms a persistent representation, or memory, for the sequence. Learning sequences is the most basic ingredient for forming invariant representations of real-world objects.

Real-world objects can be concrete, like a lizard, a face, or a door, or they can be abstract, like a word or a theory. The brain treats abstract and concrete objects in the same way. They are both just sequences of patterns that occur together over time in a predictable fashion. The fact that certain input patterns repeat time and again is what lets a cortical region know that those experiences are caused by a real object in the world.

Predictability is the very definition of reality. If a region of cortex finds it can reliably and predictably move among these input patterns using a series of physical motions (such as saccades of the eyes or fondling with the fingers) and can predict them accurately as they unfold in time (such as the sounds comprising a song or a spoken word), the brain interprets these as having a causal relationship. The odds of numerous input patterns occurring in the same relation over and over again by sheer coincidence are vanishingly small. A predictable sequence of patterns must be part of a larger object that really exists. So reliable predictability is an ironclad way of knowing that different events in the world are physically tied together. Every face has eyes, ears, mouth, and nose. If the brain sees an eye, then saccades and sees another eye, then saccades and sees a mouth, it can feel certain it is seeing a face.

If cortical regions could speak they might say, "I experience many different patterns. Sometimes I can't predict what pattern I will see next. But these patterns are definitely related to one another. They always occur together, and I can reliably jump between them. So whenever I see any of these events, I will refer to them by a common name. It is this group name, not the individual patterns, that I will pass on to higher regions of the cortex."

Therefore, the brain can be said to store sequences of sequences. Each region of the cortex learns sequences, develops what I call "names" for the sequences it knows, and passes these names to the next regions higher in the cortical hierarchy.

SEQUENCES OF SEQUENCES

As information moves up from primary sensory regions to higher levels, we see fewer and fewer changes over time. In primary visual areas like V1, the set of active cells is changing rapidly as new patterns fall on the retina several times each second. In visual area IT, cell firing patterns are more stable. What is happening here? Each region of cortex has a repertoire of sequences it knows, analogous to a repertoire of songs. Regions store these songlike sequences about anything and everything: the sound of surf crashing on the beach, your mother's face, the path from your home to the corner store, how to spell the word "popcorn," how to shuffle a deck of cards.

We have names for songs, and in a similar fashion each cortical region has a name for each sequence it knows. This "name" is a group of cells whose collective firing represents the set of objects in the sequence. (Never mind for now how that group of cells is selected to represent the sequence; we will get to this later.) These cells remain active as long as the sequence is playing, and it is this "name" that is passed up to the next region in the hierarchy. As long as the input patterns are part of a predictable sequence, the region presents a constant "name" to the next higher region.

It's as if the region were saying, "Here is the name of the sequence that I am hearing, seeing, or touching. You don't need to know about the individual notes, edges, or texture. I will let you know if something new or unpredicted happens." More specifically, we can imagine region IT at the top of the visual hierarchy relaying to an association area above it, "I am seeing a face. Yes, with each saccade the eyes are fixating on different parts of the face; I am seeing different parts of the face in succession. But it is still the same face. I will let you know when I see something else." In this fashion, a predictable sequence of events gets identified with a "name"—a constant pattern of cell firing. This happens over and over again as we go up the hierarchical pyramid. One region might recognize a sequence of sounds that comprise phonemes (the sounds that make up words) and passes a pattern representing the phoneme up to the next region. The next higher region recognizes sequences of phonemes to create words. The next higher region recognizes sequences of words to create phrases, and so on. Bear in mind that a "sequence" in the lowest regions of cortex may be fairly simple, such as a visual edge moving through space.

By collapsing predictable sequences into "named objects" at each region in our hierarchy, we achieve more and more stability the higher we go. This creates invariant representations.

The opposite effect happens as a pattern moves back down the hierarchy: stable patterns get "unfolded" into sequences. Let's assume you memorized the Gettysburg Address when you were in the seventh grade and now you want to recite it. In a language region high up in your cortex, there is a stored pattern that represents Lincoln's famous speech. First, this pattern is unfolded into a memory of the sequence of phrases. In the next region down, each phrase is unfolded into a memory of the sequence of words. At this point, the unfolding pattern splits and travels down both the auditory section of cortex and the motor section of cortex. Following the motor path, each word is unfolded into a memorized sequence of phonemes. And in the final, bottom region, each phoneme is unfolded into a sequence of muscle commands to make sounds. The lower you look in the hierarchy, the faster the patterns are changing. A single, constant pattern at the top of your motor hierarchy eventually leads to a complex and lengthy sequence of speech sounds.

Invariance also works to our advantage as this information flows back down the hierarchy. If you want to type the Gettysburg Address instead of speak it, you start out with the same pattern at the top of your hierarchy. The pattern is unfolded into phrases in the next region down. The phrases are unfolded into words in the region below that. So far there is no difference between speaking and typing the Gettysburg Address. But at the next level down, your motor cortex takes a different path. Words are unfolded into letters, and the letters are unfolded into muscle commands to your fingers for typing. "Four score and seven years ago our fathers brought forth . . ." The memories of the words are handled as invariant representations; it doesn't matter whether you will speak, type, or handwrite them. Notice you don't have to memorize the speech twice, once for speaking and once for writing. A single memory of the speech can take various behavioral forms. In any region, an invariant pattern can bifurcate and follow a different path down.

In a complementary bit of efficiency, representations of simple objects at the bottom of the hierarchy can be reused over and over for different high-level sequences. For instance, we don't have to learn one set of words for the Gettysburg Address and a completely different set for Martin Luther King's "I Have a Dream" speech, even though the two orations contain some of the same words. A hierarchy of nested sequences allows the sharing and reuse of lower-level objects—words, phonemes, and letters being but a few examples. It is a remarkably efficient way to store information about the world and its structure and very different from how computers work.

The same unfolding of sequences occurs in the sensory as well as the motor regions. The process allows you to perceive and understand objects from different views. If you are walking up to your refrigerator to get some ice cream, your visual cortex is active on multiple levels. At a higher level, you are perceiving a constant "refrigerator." In lower regions, this visual expectation is broken down into a series of more localized visual inputs. Seeing the refrigerator is composed of fixations on the door handle, the ice dispenser, the magnets on the door, a child's drawing, and so forth. In the few milliseconds that elapse as you saccade from one feature of your refrigerator to another, predictions about the result of each saccade are cascading down your visual hierarchy. As long as these predictions get confirmed saccade after saccade, your higher visual regions remain satisfied that you are in fact looking at your refrigerator. Note that in this case, unlike the fixed order of words in the Gettysburg Address, the sequence you see when looking at the refrigerator is not fixed; the flow of inputs and retrieved memory patterns depend on your own actions. So in a case like this, the unfolding pattern is not a rigid sequence, but the end result is the same: slow-changing, high-level patterns unfolding into faster-changing, low-level patterns.

The way you memorize sequences and represent them by name as information goes up and down your cortical hierarchy may remind you of the hierarchy of military command. The top army general says, "Move the troops to Florida for the winter." This simple high-level command gets unfolded into ever more detailed sequences of commands as it percolates down the hierarchy. The general's underlings recognize that the command requires a sequence of steps such as preparations to leave, transportation to Florida, and preparations for arrival. Each of these steps breaks down into further, more specific steps, to be carried out by subordinates. At the bottom there are thousands of privates taking tens of thousands of actions that result in the troops moving. Reports of what happened are generated at each level. As they percolate back up the hierarchy, they are summarized again and again, until at the very top of the hierarchy, the general receives a daily briefing saying, "Move to Florida going okay." The general does not get all the details.

There is an exception to this rule. If something goes wrong that cannot be handled by subordinates down the chain of command, then the issue rises up the hierarchy until someone knows what to do next. The officer who does know how to handle the situation does not see it as an exception. What was an unanticipated problem to subordinates is just the expected next task on his list. The officer then issues new commands to subordinates. The neocortex behaves similarly. As we will see in a bit, when events (in other words, patterns) occur that aren't anticipated, information about them progresses up the cortical hierarchy until some region can handle it. If lower regions of cortex fail to predict what patterns they are seeing, they consider this an error and pass the error up the hierarchy. This is repeated until some region does anticipate the pattern.

By design, every cortical region attempts to store and recall sequences. But this is still too simple a description of the brain. We need to add a few more complexities into the model.

The bottom-up inputs to a region of cortex are input patterns carried on thousands or millions of axons. These axons come from different regions and contain all sorts of patterns. The number of possible patterns that can exist on even one thousand axons is larger than the number of molecules in the universe. A region will only see a tiny fraction of these possible patterns in a lifetime.

So here's a question: When a single region stores sequences, what are they sequences of? The answer is that a region first classifies its inputs as one of a limited number of possibilities, and then looks for sequences. Imagine that you are a single cortical region. Your task is to sort colored pieces of paper. You are supplied ten buckets, each of which is labeled with a sample color swatch. There is one bucket for green, one for yellow, another for red, and so on. You are then given pieces of colored paper, one by one, and told to sort them by color. Each paper you receive is slightly different. Because there are an infinite number of colors in the world, you never get two pieces of paper with exactly the same color. Sometimes it is easy to say what bucket the colored paper should be placed in, but sometimes it is difficult. A paper that is halfway between red and orange could go in either bucket, but you have to assign it to one bucket, either red or orange, even if the selection comes down to a random pick. (The point of this exercise is to show that the brain must classify patterns. Regions of cortex do this, but there is nothing equivalent to a bucket to put patterns in.)

Now you are given the additional task of looking for sequences. You notice that red-red-green-purple-orange-green occurs frequently. You call it the "rrgpog" sequence. Notice that recognizing any sequence would be impossible if you hadn't first classified each piece of paper. Without first classifying each piece of paper into one of ten categories, you couldn't say that two sequences are the same.

So now you are up and running. You are going to look at all the input patterns—the colored pieces of paper coming in from lower cortical regions—classify them, and then look for sequences. Both steps, classification and sequence formation, are necessary to create invariant representations, and each region of cortex does them.

The process of forming sequences pays off when an input is ambiguous, like a piece of paper that falls somewhere between red and orange. You have to pick a bucket for the paper even if you're not sure if it's more red or more orange. If you know the most likely sequence for this series of inputs, you will use this knowledge to decide how to classify the ambiguous input. If you believe you are in the "rrgpog" sequence because you have just gotten two reds, a green, and a purple, you're going to expect that the next paper will be orange. But the next piece of paper arrives and it's not orange. Rather, it is an odd color somewhere between red and orange. It might even be a little more red than orange. But you are familiar with and expecting the "rrgpog" sequence, and so you place the paper in the orange bucket. You use the context of known sequences to resolve ambiguity.

We see this phenomenon happening all the time in our everyday experiences. When people speak, their individual words very often cannot be understood out of context. Yet when you hear an ambiguous word in a sentence, you don't get hung up on the word's ambiguity. You understand it. Similarly, handwritten words are often unintelligible out of context, but very readable within a full written sentence. Most of the time you aren't aware that you are filling in ambiguous or incomplete information from your memories of sequences. You hear what you expect to hear and see what you expect to see—at least when what you hear and see fits into past experience.

Notice the memory of sequences allows you not only to resolve ambiguity in the current input, but also to predict which input should happen next. While your cortical self is sorting colored papers, you can tell the "input" person passing the papers to you, "Hey, in case you are having any trouble deciding what to pass me next, it should, according to my memory, be an orange one." By recognizing a sequence of patterns, a cortical region will predict its next input pattern and tell the region below what to expect.

A region of cortex not only learns familiar sequences, it also learns how to modify its classifications. Let's say you start out with a set of buckets labeled "green," "yellow," "red," "purple," and "orange." You are prepared to recognize the "rrgpog" sequence as well as other combinations of these colors. But what if a color makes a major shift? What if every time you see the "rrgpog" sequence, the purple is way off-kilter? The new color is more like indigo. So you change the purple bucket to be the "indigo" bucket. Now the buckets better fit what you see; you have reduced the ambiguity. The cortex is flexible.

In cortical regions, bottom-up classifications and top-down sequences are constantly interacting, changing throughout your life. This is the essence of learning. In fact, all regions of the cortex are plastic, thus they can be modified by experience. Forming new classifications and new sequences is how you remember the world.

Last, let's look at how these classifications and predictions interact with the next higher region. Another part of your cortical job is to relay the name of the sequence you are seeing to the next level up, so you pass up a piece of paper with the letters "rrgpog" on it. These letters mean little in themselves to the next higher region; the name is just a pattern to be combined with other inputs, classified, and then put into yet a higher-order sequence. Like you, he is keeping track of the sequences he is seeing. At some point he might say to you, "Hey, in case you are having any trouble deciding what to pass me next, according to my memory, I predict it should be the 'yyrgy' sequence." This is basically an instruction to you about what to look for in your own input stream. You will do your best to interpret what you see as comprising that sequence.

Since many people have heard the term *pattern classification* used in AI and machine vision research, let's look at how this process as it is usually understood differs from what the cortex is doing. In trying to get machines to recognize objects, researchers typically create a template—say the image of a cup, or some prototypical form of cup—and then instruct the machine to match its inputs with the prototypical cup. If it finds a close match, the computer will say it has found a cup. But our brains don't have templates like that, and the patterns that each cortical region receives as input aren't like pictures. You don't remember snapshots of what your retina sees, or snapshots of the patterns from your cochlea or your skin. The hierarchy of the cortex ensures that memories of objects are distributed over the hierarchy; they aren't located in a single spot. Also, because each region of the hierarchy forms invariant memories, what a typical region of cortex learns is sequences of invariant representations, which are themselves sequences of invariant memories. You won't find a picture of a cup or any other object stored in your brain.

Unlike a camera's memory, your brain remembers the world as it is, not as it appears. When you think about the world, you are recalling sequences of patterns that correspond to the way the objects in the world are and how they behave, not how they appear through any particular sense at any point in time. The sequences by which you experience objects in the world reflect the invariant structure of the world itself. The order in which you experience parts of the world is determined by the world's structure. For example, you can get onto an airplane directly by walking down a jetway, but not from the ticket counter. The sequences by which you experience the world *is* the real structure of the world, and that is what the cortex wants to remember.

Don't forget, though, an invariant representation in any region of the cortex can be turned into a detailed prediction of how it will appear on your senses by propagating the pattern down the hierarchy. Similarly, an invariant representation in the motor cortex can be turned into detailed and situation-specific motor commands by propagating the pattern down the motor hierarchy.

WHAT A REGION OF CORTEX LOOKS LIKE

We are now going to turn our attention to an individual region of cortex, one of the boxes in figure 5. Figure 6 shows such a

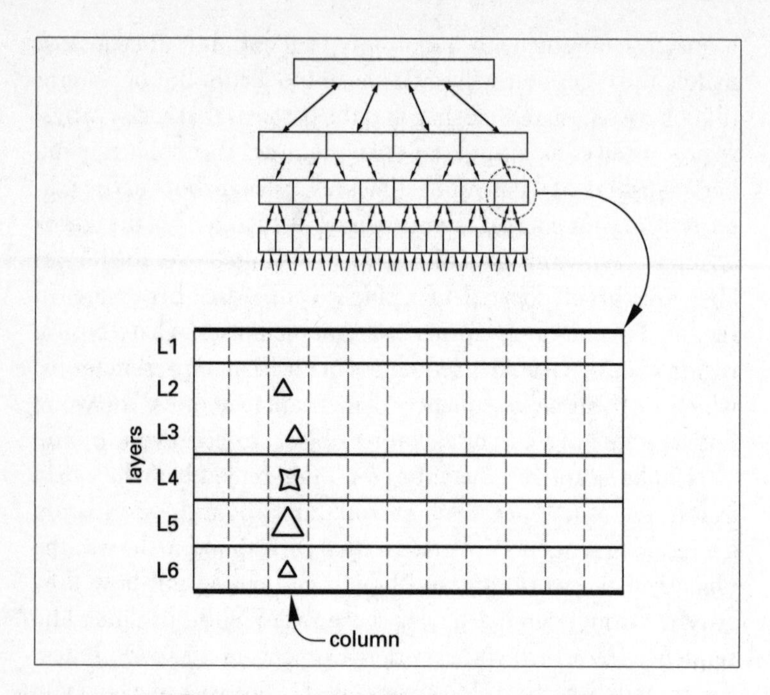

Figure 6. Layers and columns in a region of cortex.

region of cortex in more detail. My goal is to show you how the cells in a region of cortex can learn and recall sequences of patterns, which is the most essential element for forming invariant representations and making predictions. We will start with a description of what a cortical region looks like, and how it is put together. Cortical regions vary greatly in size, the largest being the primary sensory areas. V1, for example, is roughly the size of a passport in terms of the space it occupies at the back of the brain. But as I argued earlier, it is actually composed of many smaller regions that might be the size of the letters on this page. For now, let's assume that a typical cortical area is the size of a small coin.

Think of the six business cards I mentioned in chapter 3, where each card represents a different layer of cortical tissue. Why do we say there are layers? If you take our coin-size region of cortex and place it under a microscope, you'll see that the density and shape of the cells vary as you move from top to bottom. These differences define the layers. The top, called layer 1, is the most distinct of the six layers. It has very few cells, consisting primarily of a mat of axons running parallel to the cortical surface. Layers 2 and 3 look similar. They contain many tightly packed pyramidal cells. Layer 4 has a type of star-shaped cell. Layer 5 has regular pyramidal cells as well as a class of extra-big pyramid-shaped cells. The bottom layer, layer 6, also has several types of unique neurons.

Visually we see horizontal layers, but most often scientists talk about columns of cells that run perpendicular to the layers. You can think of columns as being vertical "units" of cells that work together. (The term *column* invites much debate in the neuroscience community. Their size, function, and importance are disputed. For our purposes, though, you can think in general terms of a columnar architecture, which everyone agrees exists.) The layers within each column are connected via axons that run up and down, making synapses along the way. Columns do not stand out like neat little pillars with clear boundaries—nothing in the cortex is that simple—but their existence can be inferred from several lines of evidence.

One reason is that the vertically aligned cells in each column tend to become active for the same stimulus. If we looked closely at columns in V1, we'd find some that respond to line segments that tilt in one direction (/) and some that respond to line segments that tilt in another direction (\). The cells within each column are strongly connected, which is why the entire column responds to the same stimulus. Specifically, an active cell in layer 4 causes cells above it in layers 3 and 2 to become active, which then cause cells below in layers 5 and 6 to become active. Activity spreads up and down within a column of cells.

Another reason we talk of columns stems from how the cortex forms. In an embryo, single precursor cells migrate from an inner brain cavity to where the cortex takes shape. Each of these cells divides to create about one hundred neurons, called a microcolumn, which are connected in the vertical fashion I just described. The term *column* is often used loosely to describe different phenomena; it can refer to general vertical connectivity or to specific groups of cells from the same progenitor. Using the latter definition, we can say that the human cortex has an estimated several hundred million microcolumns.

To help you visualize this columnar structure, imagine a single microcolumn is the width of a human hair. Take thousands of hairs and cut them into very short segments—say, the height of a lowercase *i* without the dot. Line up all these hairs or columns and glue them side by side like a very dense brush. Then create a sheet of long, extra-thin hairs—representing your layer 1 axons—and glue them horizontally across the top of the mat of short hairs. This brushlike mat is a simplistic model of your coin-size cortical region. Information flows mostly in the direction of these hairs: horizontally in layer 1 and vertically in layers 2 through 6.

There's one more detail about columns you need to know, and then we'll move on to what they're good for. On close inspection, we see that at least 90 percent of the synapses on cells within each column come from places outside the column itself. Some connections arrive from neighboring columns. Others come from halfway across the brain. How, then, can we speak of columns as being important when so much of our cortical wiring is spread out laterally over large areas?

The answer is in the memory-prediction model. In 1979, when Vernon Mountcastle argued that there is a single cortical algorithm, he also proposed that the cortical column is the basic unit of computation in the cortex. However, he did not know what function a column performs. I believe that a column is the basic unit of prediction. For a column to predict when it should be active, it needs to know what is going on elsewhere—hence the synaptic connections from hither and yon.

We'll get into more details soon, but here's a preview of why we need this sort of wiring in the brain. To predict the next note of a song, you need to know the song's name, where you are in the song, how much time has passed since the last note, and what that last note was. The large number of synapses connecting cells in a column to other parts of the brain provide each column with the context it needs in order to predict its activity in many different situations.

The next thing we need to consider is how these coin-size cortical regions (and their columns) send and receive information up and down the cortical hierarchy. We'll look at the upward flow first, which takes a relatively direct route, depicted in figure 7. Imagine we're looking at a cortical region with its thousands of columns. We zoom in on just one. Converging inputs from lower regions always arrive at layer 4—the main input layer. In passing, they also form a connection in layer 6 (we'll see later why this is important). Layer 4 cells then send projections up to cells in layers 2 and 3 within their column. When a column projects information up, many layer 2 and layer 3 cells send axons to the input layer of that next higher region. Thus information flows from region to region up the hierarchy.

Information flowing down the cortical hierarchy takes a less direct path, as depicted in figure 8. Layer 6 cells are the downward-projecting output cells from a cortical column and project to layer 1 in the regions hierarchically below. Here in

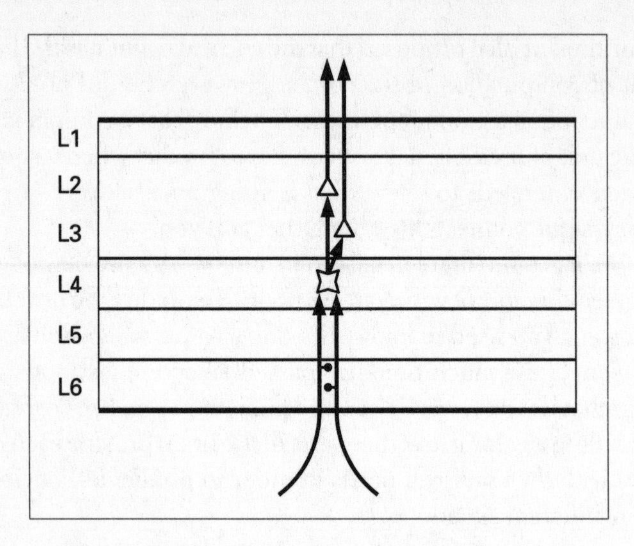

Figure 7. Upward flow of information through a region of cortex.

---

layer 1, the axon spreads over long distances in the lower cortical region. Thus information flowing down the hierarchy from one column has the potential to activate many columns in the regions below it. There are very few cells in layer 1, but cells in layers 2, 3, and 5 have dendrites in layer 1, so these cells can be excited by the feedback running all across layer 1. The axons coming from layers 2 and 3 cells form synapses in layer 5 as they leave the cortex and are believed to excite cells in layers 5 and 6. So we can say that as information flows down the hierarchy, it has a less direct route. It can branch in many different directions via the spread on layer 1. Feedback information starts in a layer 6 cell in the higher region; it spreads across layer 1 in the lower region. Some cells in layers 2, 3, and 5 in the lower region are excited, and some of these excite layer 6 cells, which project to layer 1 in regions hierarchically below, and so on. (If you study figure 8, the process is much easier to follow.)

Here is a preview of why information is spread across layer 1.

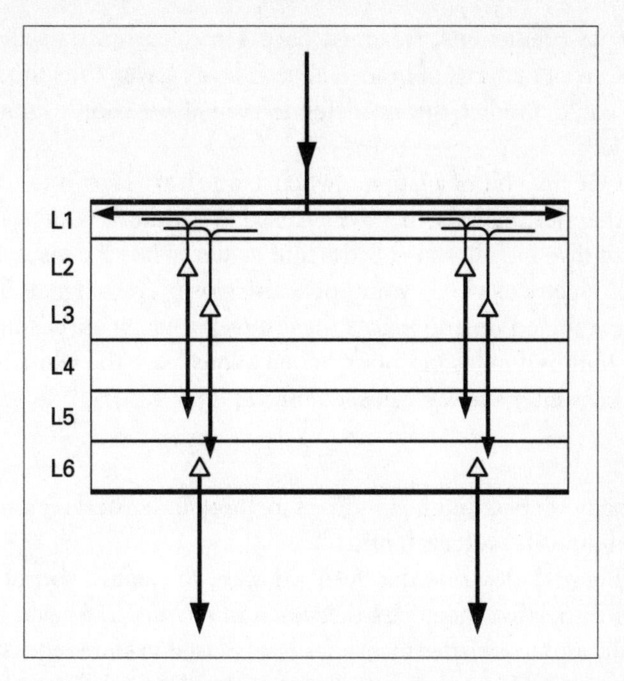

Figure 8. Downward flow of information through a region of cortex.

To convert an invariant representation into a specific prediction requires the ability to decide moment by moment which way to send the signal as it propagates down the hierarchy. Layer 1 provides a way of converting an invariant representation into a more detailed and specific representation. Remember that you can recall the Gettysburg Address either in spoken language or in writing. A common representation is moved along one of two paths, one for speaking and one for writing. Similarly when I hear the next note in a melody, my brain has to take a generic interval such as a fifth and convert it to the correct specific note, such as C or G. The horizontal flow of activity across layer 1 provides the mechanism for doing this. For high-level invariant predictions to propagate down the cortex and become specific predictions, we must have a mechanism that allows the flow of patterns to branch at each level. Layer 1 fits the bill. We could predict the need for it even if we didn't know it existed.

One final bit of anatomy: when axons leave layer 6 to travel to other destinations, they are encased in a white fatty substance called myelin. This so-called white matter is like the insulation on an electrical wire in your house. It helps prevent signals from getting mixed up and makes them travel faster, at speeds up to two hundred miles per hour. When axons leave the white matter, they enter a new cortical column at layer 6.

Finally, there is another, indirect method for cortical regions to communicate with each other.

Before I describe the details, I want to remind you of the auto-associative memories discussed in chapter 2. As you may recall, auto-associative memories can be used to store sequences of patterns. When the output of a group of artificial neurons is fed back to form the input to all the neurons, and a delay is added to the feedback, then patterns learn to follow each other in sequence. I believe the cortex uses the same basic mechanism to store sequences, although with a few extra twists. Instead of forming an auto-associative memory out of artificial neurons, it is forming an auto-associative memory out of cortical columns. The output of all the columns is fed back to layer 1. In this way, layer 1 contains information about which columns were just active in the region of cortex.

Let's walk through the elements, as shown in figure 9. It has been known for many years that particularly large layer 5 cells within your motor cortex (region M1) make direct contact with your muscles and the motor regions of your spinal cord. These cells literally drive your muscles and make you move. Whenever

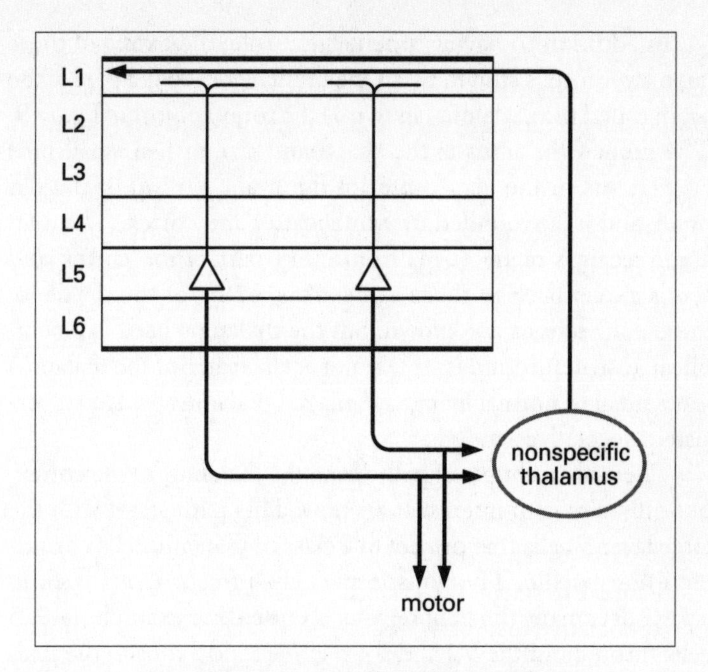

Figure 9. How current state and current motor behavior is communicated broadly via the thalamus.

you speak, type, or perform any sophisticated behavior, these cells are firing on and off in a highly coordinated way, making your muscles contract.

Recently, researchers have discovered that large layer 5 cells may play a role in behavior in other parts of the cortex, not just in motor regions. For example, large layer 5 cells in visual cortex project to the part of the brain that moves the eyes. So the sensory visual areas of cortex, such as V2 and V4, not only process visual input, but they help determine the movement of the eyes themselves, and therefore what you see. Large layer 5 cells are seen throughout the neocortex, in every region, suggesting a more widespread role in all kinds of movement.

In addition to having a behavioral role, the axons of these large layer 5 cells split in two. One branch goes to the part of the brain called the thalamus, shown as the round object in figure 9. The human thalamus is the shape and size of two small bird eggs. It sits in the very center of the brain, on top of the old brain, and is surrounded by white matter and cortex. The thalamus receives many axons from every part of the cortex and sends axons back to those same areas. Many of the details of these connections are known, but the thalamus itself is a complicated structure and its role is not at all clear. But the thalamus is essential to normal living; a damaged thalamus leads to a persistent vegetative state.

There are a couple of paths from the thalamus to the cortex, but only one is of interest to us now. This path starts with the large layer 5 cells that project to a class of thalamic cells considered nonspecific. The nonspecific cells project axons back to layer 1 over many different regions of cortex. For example, layer 5 cells throughout the V2 and V4 regions send axons to the thalamus, and the thalamus sends the information back to layer 1 throughout V2 and V4. Other parts of the cortex do the same; layer 5 cells throughout multiple cortical regions project to the thalamus, which sends the information back to layer 1 of those same and associated regions. I propose this circuit is exactly like the delayed feedback that lets auto-associative memory models learn sequences.

I have now mentioned two inputs to layer 1. Higher regions of cortex spread activity across layer 1 in lower regions. Active columns within a region also spread activity across layer 1 in the same region via the thalamus. We can think of these inputs to layer 1 as the name of a song (input from above) and where we are in a song (delayed activity from active columns in the same region). Thus layer 1 carries much of the information we need to predict when a column should be active—the sequence name and where we are within the sequence. Using these two signals in layer 1, a region of cortex can learn and recall multiple sequences of patterns.

HOW A REGION OF CORTEX WORKS: THE DETAILS

With these three circuits in mind—converging patterns going up the cortical hierarchy, diverging patterns going down the cortical hierarchy, and a delayed feedback through the thalamus—we can start to see how a region of cortex performs the functions it needs. What we want to know is:

1. How does a region of cortex classify its inputs (like the buckets)?
2. How does it learn sequences of patterns (such as intervals of a melody, or the "eye nose eye" of a face)?
3. How does it form a constant pattern or "name" for a sequence?
4. How does it make specific predictions (meeting the train at the right time, or predicting the specific note in a melody)?

Let's start by assuming that the columns in a region of cortex are like the buckets we used in classifying our colored paper inputs. Each column represents the label of a bucket. The layer 4 cells in every column receive input fibers from several regions below it and will fire if they have the right combination of inputs. When a layer 4 cell fires, it is "voting" that the input fits its label. As in the paper-sorting analogy, inputs can be ambiguous, so several columns might be possible matches for the input. We want our region of cortex to decide on one interpretation; the paper is either red or orange, but not both. A column with strong input should prevent other columns from firing.

Brains have inhibitory cells that do just this. They strongly inhibit other neurons in a neighborhood of cortex, effectively allowing one winner. These inhibitory cells only affect the area surrounding a column. So even though there is a lot of inhibition, many columns in a region can still be active at once. (In real brains, nothing is ever represented by a single neuron or a single column.) To make it easier to follow, you can pretend that a region picks one and only one winner column. But in the back of your mind, remember that many columns will be active at once. The actual process used by a region of cortex to classify inputs and how it learns to do this is complicated, and not well understood. I will not attempt to drag you through the issues. Instead I want to assume that our region of cortex has classified its input as activity in a set of columns. We can then focus on the forming of sequences and names for sequences.

How does our cortical region store the sequence of these classified patterns? I have already suggested an answer to this question, but now I'll delve into more details. Imagine you are a column of cells, and input from a lower region causes one of your layer 4 cells to fire. You are happy, and your layer 4 cell causes cells in layers 2 and 3, then 5, and then 6 also to fire. The entire column becomes active when driven from below. Your cells in layers 2, 3, and 5 each have thousands of synapses in layer 1. If some of these synapses are active when your layer 2, 3, and 5 cells fire, the synapses are strengthened. If this occurs often enough, these layer 1 synapses become strong enough to make the cells in layers 2, 3, and 5 fire even when a layer 4 cell hasn't fired—meaning parts of the column can become active without receiving input from a lower region of the cortex. In this way, cells in layers 2, 3, and 5 learn to "anticipate" when they should fire based on the pattern in layer 1. Before learning, the column can only become active if driven by a layer 4 cell. After learning, the column can become partially active via memory. When a column becomes active via layer 1 synapses, it is anticipating being driven from below. This is prediction. If the column could speak, it would say, "When I have been active in the past, this particular set of my layer 1 synapses has been active. So when I see this particular set again, I will fire in anticipation."

Recall that half of the input to layer 1 comes from layer 5 cells in neighboring columns and regions of the cortex. This information represents what was happening moments before. It represents columns that were active prior to your column becoming active. It represents the previous interval in the melody, or the last thing I saw, or the last thing I felt, or the previous phoneme in the speech I am listening to. If the order in which these patterns occur over time is consistent, then the columns will learn the order. They will fire one after another in proper sequence.

The other half of the input to layer 1 comes from layer 6 cells in hierarchically higher regions. This information is more stationary. It represents the name of the sequence you are currently experiencing. If your columns are music intervals, it is the name of the melody. If your columns are phonemes, then it is the spoken word you are hearing. If your columns are spoken words, then the signal from above is the speech you are reciting. Thus the information in layer 1 represents both the name of a sequence and the last item in the sequence. In this way, a particular column can be shared among many different sequences without getting confused. Columns learn to fire in the right context and in the correct order.

Before moving on, I need to point out that the synapses in layer 1 are not the only synapses that participate in learning when a column should become active. As I mentioned earlier, cells receive input from, and send input to, many surrounding columns. Recall that more than 90 percent of all synapses are from cells outside the column, and most of these synapses are not in layer 1. For example, cells in layers 2, 3, and 5 have thousands of synapses in layer 1, but also thousands of synapses in their own layers. The overall idea is that cells want any information that will help them predict when they will be driven from below.

Usually, activity in nearby columns has a strong correlation, and so we see many direct connections to nearby columns. For example, if a line is moving through your visual field, it will activate successive columns. Often, though, the information needed to predict a column's activity is more global, which is where the layer 1 synapses play a role. If you were a cell or a column, you wouldn't know what any of these synapses meant, all you would know is that they help you predict when you should become active.

Now let us consider the issue of how a region of cortex forms a name for a learned sequence. Again, imagine you are a region of cortex. Your active columns are changing with each new input. You have successfully learned the order in which your columns become active, meaning some of the cells in your columns become active prior to the arrival of inputs from lower regions. What information are you sending to the hierarchically higher regions of the cortex? We saw earlier that your layer 2 and layer 3 cells send their axons to the next higher regions. The activity of these cells is the input to the higher regions. But that's a problem. In order for the hierarchy to work, you have to relay a constant pattern during learned sequences; you have to pass on the name of a sequence, not the details. Before you learn a sequence, you can pass on the details, but after you learn a sequence, and are able to successfully predict which columns will be active, you should relay only a constant pattern. However, I have not yet shown you a method for doing this. As it currently stands, you will pass on every changing pattern regardless of whether you can predict it. As each column becomes active, its layer 2 and layer 3 cells will send a new signal up the hierarchy. The cortex needs some way to keep the input to the next region constant during learned sequences. We need some way to turn off the output of the layer 2 and layer 3 cells when a column predicts its activity, or, alter-

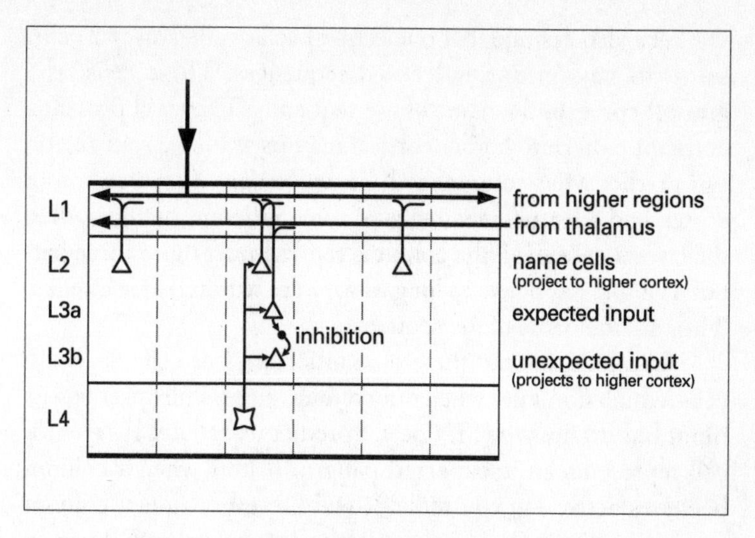

Figure 10. Forming a constant name for a learned sequence.

nately, to make these cells active when the column can't predict its activity. This is the only way to make a constant name pattern.

There is not enough known about the cortex to state exactly how it does this. I can imagine several methods. I will describe my current favorite, but keep in mind that the concept is more important than the specific method. The creation of a constant "name" pattern is a requirement of this theory. All I can show at this time is that plausible mechanisms exist for the naming process.

Again imagine you are a column, as shown in figure 10. We want to understand how you learn to present a constant pattern to the next higher region when you can predict your activity, and a changing pattern when you can't predict your activity. Let's start by assuming that within layers 2 and 3 there are several classes of cells. (In addition to several types of inhibitory cells, many anatomists make the distinction between cell types in what they call layers 3a and 3b, so this assumption is not unreasonable.)

Let's also assume that one class of cells, called layer 2 cells, learns to stay on during learned sequences. These cells, as a group, represent the name of the sequence. They will present a constant pattern to higher cortical regions as long as our region can predict what columns will be active next. If our region of cortex had a learned sequence of three different patterns, then the layer 2 cells of all the columns representing those three patterns would stay active as long as we were within that sequence. They are the name of the sequence.

Next, let's assume there is another class of cells, layer 3b cells, which don't fire when our column successfully predicts its input but do fire when it doesn't predict its activity. A layer 3b cell represents an unexpected pattern. It fires when a column becomes active unexpectedly. It will fire every time a column becomes active prior to any learning. But as a column learns to predict its activity, the layer 3b cell becomes quiet. Together, the layer 2 and layer 3b cells meet our requirement. Before learning both cells fire on and off with the column, but after training the layer 2 cell is constantly active and the layer 3b cell is quiet.

How do these cells learn to do this? First, let's consider how to shut down the layer 3b cell when its column successfully predicts its activity. Say that there is another cell positioned just above the layer 3b cell, in layer 3a. This cell also has dendrites in layer 1. Its only job is to prevent the layer 3b cell from firing when it sees the appropriate pattern in layer 1. When the layer 3a cell sees the learned pattern in layer 1, it quickly activates an inhibitory cell that prevents the layer 3b cell from firing. This is all it would take to stop the layer 3b cell from firing when the column correctly predicts its activity.

Now consider the more difficult task of keeping the layer 2 cell active throughout a known sequence of patterns. This is more difficult because a diverse set of layer 2 cells in many different columns would need to stay active together, even when their individual columns are not active. Here is how I believe this could occur. Layer 2 cells could learn to be driven purely from the hierarchically higher regions of cortex. They could form synapses preferentially with the axons from layer 6 cells in the regions above. The layer 2 cells would therefore represent the constant name pattern from the higher region. When a higher region of cortex sends a pattern down to layer 1 of the region below, a set of layer 2 cells in the lower region would become active, representing all the columns that are members of the sequence. Since these layer 2 cells also project back to the higher region, they would form a semistable group of cells. (It is unlikely that these cells stay constantly active. They probably fire synchronously in something like a rhythm.) It is as if the higher region sends the name of a melody to layer 1 below. This event causes a set of layer 2 cells to fire, one for each of the columns that will be active as the melody is heard.

The sum of all these mechanisms allows the cortex to learn sequences, make predictions, and form constant representations, or "names," for sequences. These are the basic operations for forming invariant representations.

How do we make predictions about events we have never seen before? How do we decide among multiple interpretations of an input? How does a region of the cortex make specific predictions from invariant memories? I gave several examples of this earlier, such as predicting the exact next note in a melody when your memory only recalls the interval between the notes, the parable of the train, and reciting the Gettysburg Address. In these cases, the only way to solve this problem is to use the last specific information to convert an invariant prediction into a specific prediction. Another way to phrase this, in terms of the cortex, is to say that we have to combine feedforward information (actual input) with feedback information (a prediction in an invariant form).

Here is a simple example of how I believe this is done. Say

your region of the cortex has been told to expect the musical interval of a fifth. The columns of your region represent all possible specific intervals such as C-E, C-G, D-A, etc. You need to decide which of your columns should be active. When the region above tells you to expect a fifth, it causes layer 2 cells to fire in all the columns that are fifths, such as C-G, D-A, and E-B. Cells in layer 2 of columns representing other intervals are not active. Now you have to select one column from all the possible fifths. The inputs to your region are specific notes. If the last note you heard was a D, then all the columns representing intervals involving a D, such as D-E and D-B, have partial input. So now in layer 2 we have activity in all columns that are fifths, and in layer 4 we have partial input to all columns representing intervals involving a D. The intersection of these two sets represents our answer, the column representing the interval D-A (see figure 11).

How does the cortex find this intersection? Recall that earlier I mentioned the fact that axons from layer 2 and layer 3 cells generally form synapses in layer 5 as they leave the cortex, and similarly axons approaching layer 4 from lower regions of cortex make a synapse in layer 6. The intersection of these two synapses (top down and bottom up) provides us with what is needed. A layer 6 cell that receives these two active inputs will fire. A layer 6 cell represents what a region of cortex believes is happening, a specific prediction. If a layer 6 cell could talk, it might say, "I am part of a column representing something. In my particular case my column represents the musical interval D-A. Other columns mean other things. I speak for my region of cortex. When I become active it means we believe the musical interval D-A either is occurring or will occur. I might become active because the bottom-up input from the ears caused the layer 4 cell in my column to excite the entire column. Or my activity might mean that we recognized a melody and are predicting this specific next interval. Either way, my job is to tell the lower regions of cortex what we think is happening. I represent

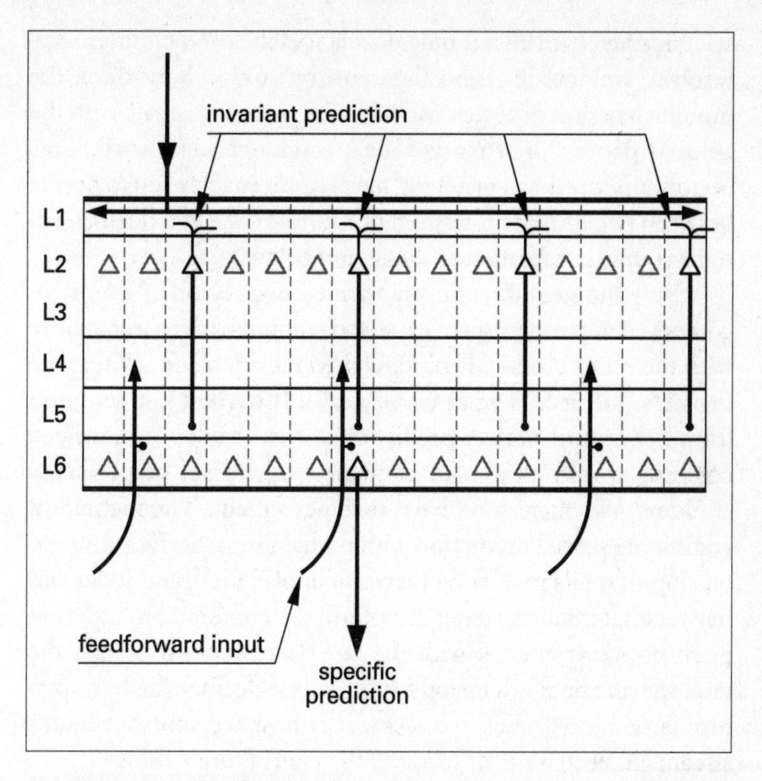

Figure 11. How a region of cortex makes specific predictions from invariant memories.

---

our interpretation of the world, regardless of whether it is true or just imagined."

Let me describe this using another mental picture. Imagine two pieces of paper with lots of little holes punched in them. The holes on one paper represent the columns that have active layer 2 or layer 3 cells, our invariant prediction. The holes on the other paper represent columns with partial input from below. If you put one piece of paper on top of the other, some of the holes will line up, others won't. The holes that line up represent the columns we think should be active.

This mechanism not only makes specific predictions, it also resolves ambiguities from the sensory inputs. Very often the input to a region of cortex will be ambiguous, as we saw with the colored papers, or when you hear a semi-garbled word. This bottom-up/top-down matching mechanism enables you to decide between two or more interpretations. And once you decide, you relay your interpretation to the region below.

Every moment in your waking life, each region of your neocortex is comparing a set of expected columns driven from above with the set of observed columns driven from below. Where the two sets intersect is what we perceive. If we had perfect input from below and perfect predictions, then the set of perceived columns would always be contained in the set of predicted columns. We often don't have such agreement. The method of combining partial prediction with partial input resolves ambiguous input, it fills in missing pieces of information, and it decides between alternative views. It is how we combine an expected pitch-invariant interval with the last heard note to predict the next specific note in a melody. It is how we decide whether a picture is of a vase or of two faces. It is how we split our motor stream either to write or to speak the Gettysburg Address.

Finally, in addition to projecting to lower cortical regions, layer 6 cells can send their output back into layer 4 cells of their own column. When they do, our predictions become the input. This is what we do when daydreaming or thinking. It allows us to see the consequences of our own predictions. We do this many hours a day as we plan the future, rehearse speeches, and worry about events to come. Longtime cortical modeler Stephen Grossberg calls this "folded feedback." I prefer "imagining."

One last topic before we leave this section. I have pointed out several times that most often what we see, hear, or feel is highly dependent on our own actions. What we see is dependent on where our eyes saccade and how we turn our heads. What we feel is dependent on how we move our limbs and fingers. What we hear is sometimes dependent on what we say and do.

Therefore, to predict what we will sense next, we have to know what actions we are undertaking. Motor behavior and sensory perception are highly interdependent. How can we make predictions if what we sense next is largely a result of our own actions? Fortunately, there is a surprising and elegant solution to this problem, although many of the details are not understood.

The first surprising discovery is that perception and behavior are almost one and the same. As I mentioned earlier, most if not all regions of the cortex, even visual areas, participate in the creation of movement. The layer 5 cells that project to the thalamus and then to layer 1 also seem to have a motor function because they simultaneously project to motor areas of the old brain. Thus, the knowledge of "what just happened"—both sensory and motor—is available in layer 1.

The second surprising thing, and a consequence of the first, is that motor behavior must also be represented in a hierarchy of invariant representations. You generate the movements necessary to carry out a particular action by thinking of doing it in a detail-invariant form. As the motor command travels down the hierarchy, it gets translated into the complex and detailed sequences required to perform the activity you expected to do. This is happening in both "motor" cortex and "sensory" cortex, which blurs the distinction between the two. If region IT of visual cortex is perceiving "nose," the mere act of switching to the representation for "eye" will generate the saccade necessary to make this prediction a reality. The particular saccade necessary to move from seeing a nose to seeing an eye varies depending on where the face is. A close face requires a larger saccade; a more distant face requires a smaller saccade. A tilted face requires saccading at an angle different from the one for a level face. The details of the needed saccade are determined as the prediction of seeing the "eye" moves toward V1. The saccade becomes increasingly specific the farther down it goes, resulting in a saccade that lands your foveas right on target, or pretty close.

Let's look at another example. For me to physically move from my living room to my kitchen, all my brain has to do is mentally switch from the invariant representation of my living room to the invariant representation of my kitchen. This switch causes a complex unfolding of sequences. The process of generating the sequence of predictions of what I will see, feel, and hear while walking from the living room to the kitchen also generates the sequence of motor commands that makes me walk from my living room to my kitchen and move my eyes as I do so. Prediction and motor behavior work hand in hand as patterns flow down and up the cortical hierarchy. As strange as it sounds, when your own behavior is involved, your predictions not only precede sensation, they determine sensation. Thinking of going to the next pattern in a sequence causes a cascading prediction of what you should experience next. As the cascading prediction unfolds, it generates the motor commands necessary to fulfill the prediction. Thinking, predicting, and doing are all part of the same unfolding of sequences moving down the cortical hierarchy.

"Doing" by thinking, the parallel unfolding of perception and motor behavior, is the essence of what is called goal-oriented behavior. Goal-oriented behavior is the holy grail of robotics. It is built into the fabric of the cortex.

We can turn off our motor behavior, of course. I can think of seeing something without actually seeing it and I can think of going to my kitchen without actually doing so. But thinking of doing something is literally the start of how we do it.

FLOWING UP AND FLOWING DOWN

Let's step back a bit and think some more about how information moves up and down the cortical hierarchy. As you move about the world, changing inputs stream into lower regions of the cortex. Each region tries to interpret its stream of inputs as part of a known sequence of patterns. The columns try to anticipate their activity. If they can, they will pass on a stable pattern, the name of the sequence, to the next higher region. Again it is as if the region says, "I am listening to a song, here is the name. I can handle the details."

But what if an unexpected pattern arrives, an unexpected note? Or what if we see something that does not belong on a face? The unexpected pattern is automatically passed to the next higher cortical region. This happens naturally as layer 3b cells that were not part of the expected sequence fire. The higher region may be able to understand this new pattern as the next part of its own sequence. It might say, "Oh, I see a new note arrived. Maybe this is the first note of the next song on the album. It looks like it, so I predict we have gone on to the next song. Lower region, here is the name of the next song I think you should be hearing." But if such recognition does not occur, an unexpected pattern will keep propagating up the cortical hierarchy until some higher region can interpret it as part of its normal sequence of events. The higher the unexpected pattern needs to go, the more regions of the cortex get involved in resolving the unexpected input. Finally, when a region somewhere up the hierarchy thinks it can understand the unexpected event, it generates a new prediction. This new prediction propagates down the hierarchy as far as it can go. If the new prediction is not right, an error will be detected, and again it will climb up the hierarchy until some region can interpret it as part of its currently active sequence. Thus we can see that observed patterns flow up the hierarchy and predictions flow down the hierarchy. Ideally, in a world that is known and predictable, most of the up-and-down flow of patterns happens rapidly and occurs in the lower regions of the cortex. The brain quickly tries to find a part of its world model that is consistent with any unexpected input. Only then will it understand its input and know what to expect next.

If I am walking in a familiar room in my house, few errors will propagate high up my cortex. The highly learned sequences of my house can be handled in the lower sections of the visual, somatosensory, and motor hierarchy. I know the room so well I can even walk around in the dark. My familiarity with the surroundings effectively frees up most of my cortex for other tasks such as thinking about brains and writing books. However, if I was in an unknown room, especially one that was different from any I had seen before, not only would I need to look to see where I was going, but unexpected patterns would constantly be rising far up the cortical hierarchy. The more my sensory experience doesn't match learned sequences, the more the errors rise up. In this novel situation, I can no longer think about brains because most of my cortex is attending to the problems of navigating the room. This is a common experience for people who step off an airplane into a foreign country. While the roads may look like those you're accustomed to, cars may whiz down the wrong side, the money is strange, the language is incomprehensible, and learning how to find a bathroom can take all your cortical powers. Don't try to rehearse a speech while walking in a foreign land.

The sensation of sudden comprehension, the "aha!" moment, can be understood in this model. Imagine you are looking at an ambiguous picture. Filled with blobs of ink and scattered lines, it doesn't look like anything. It doesn't make sense. Confusion occurs when the cortex can't find any memory that matches with the input. Your eyes scan everywhere on the picture. New inputs race all the way up the cortical hierarchy. High-level cortex tries lots of different hypotheses but, as these predictions race down the hierarchy, each and every one conflicts with the input and the cortex is forced to try again. During this time of confusion your brain is totally occupied with understanding the picture. Finally, you make a high-level prediction

Figure 12. Do you see the dalmatian?

that is the right one. When this happens, the prediction starts at the top of the cortical hierarchy and succeeds in propagating all the way to the bottom, chunk, chunk, chunk, chunk, chunk. In less than a second, each region is given a sequence that fits the data. No more errors rise to the top. You understand the picture, you see a dalmatian dog amid the dots and scribbles (see figure 12).

CAN FEEDBACK REALLY DO THAT?

We've known for decades that connections in the cortical hierarchy are reciprocal. If region A projects to region B, then B projects to region A. There are often more axon fibers going backward than forward. But even though this description is widely accepted, the prevailing paradigm is that feedback plays a minor or "modulatory" role in the brain. The idea that a feedback signal could instantly and accurately cause a diverse set of cells in layer 2 to fire is not the prevailing view among neuroscientists.

Why would this be so? Part of the reason, as I mentioned before, is that there's no real need to be concerned with feedback if you don't accept the central role of prediction. If you think that the information flows straight through to the motor system, then why do you need feedback? Another reason to ignore feedback is that the feedback signal is spread over large areas of layer 1. Intuitively, we would expect a signal dispersed over a large area would have only a minor effect on many neurons, and indeed the brain has several such modulatory signals that don't act on specific neurons but change global attributes such as alertness.

The final reason to ignore feedback is due to how many scientists believe individual neurons work. Typical neurons have thousands or tens of thousands of synapses. Some are located far from the cell body, others are right up close to it. Synapses close to the cell body have a strong influence on cell firing. A dozen or so active synapses near the cell body can cause it to generate a spike or pulse of electrical discharge. This is known. However, the vast majority of synapses are not close to the body of the cell. They are spread far and wide out over the branchlike structure of a cell's dendrites. Since these synapses are far away from the cell's body, scientists have tended to believe that a spike arriving at one of those synapses would have a weak or almost imperceptible effect on whether the neuron generates a spike. The effect of a distant synapse would dissipate by the time it reaches the cell body.

As a general rule, information flowing up the cortical hierarchy is transferred via synapses close to cell bodies. Information going up the hierarchy is therefore more certain to pass from region to region. Also, as a general rule, feedback flowing down the cortical hierarchy does so via synapses far from the cell body. Cells in layers 2, 3, and 5 send dendrites into layer 1 and form many synapses there. Layer 1 is a mass of synapses, but they are all far from the cell bodies in layers 2, 3, and 5. Furthermore, any particular cell in layer 2, say, will only form a few, if any, synapses with any particular feedback fiber. Therefore, some scientists may object to the idea that a brief pattern in layer 1 could accurately cause a set of cells to fire in layers 2, 3, and 5. But that is precisely what the theory I have laid out requires.

The resolution to this dilemma is that neurons behave differently from the way they do in the classic model. In fact, in recent years there has been a growing group of scientists who have proposed that synapses on distant, thin dendrites can play an active and highly specific role in cell firing. In these models, these distant synapses behave differently from synapses on thicker dendrites near the cell body. For example, if there were two synapses very close to each other on a thin dendrite, they would act as a "coincidence detector." That is, if both synapses received an input spike within a small window of time, they could exert a large effect on the cell even though they are far from the cell body. They could cause the cell body to generate a spike. How the dendrites of a neuron behave is still a mystery, so I cannot say much more about them here. What is important is that the memory-prediction model of cortex requires that synapses far from the cell body be able to detect specific patterns.

In hindsight, it seems almost foolish to say that most of the thousands of synapses on a neuron play only a modulatory role. Massive feedback and massive numbers of synapses exist for a reason. Using this insight, we can say that a typical neuron has the ability to learn hundreds of precise coincidences on feedback fibers as they make synapses on thin dendrites. It means that each column in our neocortex is highly flexible in terms of which feedback patterns cause it to become active. It means that any particular feature can be precisely associated with thousands of different objects and sequences. My model requires feedback to be fast and precise. Cells need to fire when they see any number of precise coincidences on their distant dendrites. These new neuron models allow for this.

HOW THE CORTEX LEARNS

All the cells in all the layers of the cortex have synapses, and most of these synapses can be modified via experience. It is safe to say that learning and memory occur in all layers, in all columns, in all regions of cortex.

Earlier in the book I mentioned Hebbian learning, named after the Canadian neuropsychologist Donald O. Hebb. Its essence is very simple: When two neurons fire at the same time, the synapses between them get strengthened. (It is easily summarized in the phrase "Fire together, wire together.") We now know Hebb was basically right. Of course, nothing in nature is ever quite so simple, and in real brains the details are more complicated. Our nervous systems run many variations of Hebb's learning rule; for instance, some synapses change their strength in response to small variations in the timing of neural signals, some synaptic changes are short-lived, and some changes are long-lived. But Hebb was erecting only a framework for the study of learning, not a final theory, and that framework has been incredibly useful.

Hebbian learning principles can explain most of the cortical behavior I have mentioned in this chapter. Remember, it was shown back in the 1970s that auto-associative memories using the classical Hebbian learning algorithm can learn spatial patterns and sequences of patterns. The main problem was that the memories couldn't handle variation well. According to the theory proposed in this book, the cortex has gotten around this limitation partly by stacking auto-associative memories in a hierarchy and partly by using a sophisticated columnar architecture. This chapter has been almost all about the hierarchy and how it works, because the hierarchy is what makes the cortex powerful. So instead of going through in painful detail how each cell could learn this or that, I want to cover a few broad principles of learning in a hierarchy.

When you are born your cortex essentially doesn't know anything. It doesn't know about your language, your culture, your home, your town, songs, the people you will grow up with, nothing. All this information, the structure of the world, has to be learned. The two basic components of learning are forming the classifications of patterns and building sequences. These two complementary memory components interact. As one region learns sequences, the inputs it sends to the layer 4 cells in higher cortical regions change. These layer 4 cells therefore learn to form new classifications, which changes the pattern projected back to layer 1 in the lower region, which affects the sequences.

The basics of forming sequences is to group patterns together that are part of the same object. One way to do this is by grouping patterns that occur contiguously in time. If a child holds a toy in her hand and slowly moves it, her brain can safely assume that the image on her retina is of the same object moment to moment, and therefore the changing set of patterns can be grouped together. At other times you need outside instruction to help you decide which patterns belong together. To learn that apples and bananas are fruits, but carrots and celery are not, requires a teacher to guide you to group these items as fruits. Either way, your brain slowly builds sequences of patterns that belong together. But as a region of cortex builds sequences, the input to the next region changes. The input changes from representing mostly individual patterns to representing groups of patterns. The input to a region changes from notes to melodies, from letters to words, from noses to faces, and so on. Because the bottom-up inputs to a region become more "object-oriented," the higher region of cortex can now learn sequences of these higher-order objects. Where before a region built sequences of letters, it now builds sequences of words. The unexpected result of this learning process is that, during repetitive learning, representations of objects move down the cortical hierarchy. During the early years of your life, your memories of the world first form in higher regions of cortex, but as you learn they are re-formed in lower and lower parts of the cortical hierarchy. It isn't that the brain moves them; it has to relearn them over and over. (I am not suggesting that all memories start at the top of the cortex. The actual formation of memories is more complex. I believe layer 4 pattern classification starts at the bottom and moves up. But as it does, we start forming sequences that then move down. It is the *memory of sequences* I am suggesting re-form lower and lower in the cortex.) As simple representations move down, the regions at the top are able to learn more complex and subtle patterns.

You can observe the creation and downward movement of hierarchical memory by observing how a child learns. Consider how we learn to read. The first thing we learn is to recognize individual printed letters. This is a slow and difficult task requiring conscious effort. Then we move on to recognizing simple words. Again, it is difficult and slow at first, even for three-letter words. The child can read each letter in sequence and sound out the letters one after another, but it takes a fair amount of practice before the word itself is recognized as a word. After learning simple words, we struggle with complex, multisyllable words. At first, we sound out each syllable, concatenating them as we did with letters when learning simple words. After years of practice, a person can read quickly. We get to the point where we don't actually see all the individual letters but instead recognize entire words and often entire phrases at a glance. It isn't just that we are faster; we are actually recognizing words and phrases as entities. When we read an entire word at one time, do we still see the letters? Yes and no. Obviously, the retina sees the letters and therefore so do regions of V1. But the recognition of the letters is occurring fairly low in the cortical hierarchy, say in V2 or V4. By the time the signal gets to IT, the individual letters are no longer represented. What at first took the effort of your entire visual cortex—recognizing individual letters—is now occurring closer to the sensory input. As memory of simple objects like letters moves down the hierarchy, the higher regions have the ability to learn complex objects like words and phrases.

Learning to read music is another example. At first you have to concentrate on every note. With practice, you start to recognize common note sequences, then entire phrases. After much practice, it is as if you don't see most of the notes at all. The sheet music is there only to remind you of the major structure of the piece; the detailed sequences have been memorized lower down. This type of learning occurs in both motor and sensory areas.

A young brain is slower to recognize inputs and slower to make motor commands because the memories used in these tasks are higher up the cortical hierarchy. Information has to flow all the way up and down, maybe with multiple passes, to resolve conflicts. It takes time for the neural signals to travel up and down the cortical hierarchy. A young brain also has not yet formed complex sequences at the top and therefore cannot recognize and play back complex patterns. A young brain cannot understand the higher-order structure of the world. Compared to an adult's, a child's language is simple, his music is simple, and his social interactions are simple.

If you study a particular set of objects over and over, your cortex re-forms memory representations for those objects down the hierarchy. This frees up the top for learning more subtle, more complex relationships. According to the theory, this is what makes an expert.

In my work designing computers, some people are surprised by how quickly I can look at a product and see the problems inherent in its design. After twenty-five years of designing computers, I have a better-than-average model of the issues associated with mobile computing devices. Similarly, an experienced parent can easily recognize why his child is upset, whereas a first-time parent may struggle with how to handle a situation. An experienced business manager can readily see the flaws and advantages of the structure of an organization whereas the novice manager just can't understand these things. They have the same input, but the novice's model is not as sophisticated. In all such cases and a thousand more, we start by learning the basics, the simplest structure. Over time we move our knowledge down the cortical hierarchy and, therefore, we have the opportunity at the top for learning higher-order structure. It is this higher-order structure that makes us experienced. Experts and geniuses have brains that see structure of structure and patterns of patterns beyond what others do. You can become expert by practice, but there certainly is a genetic component to talent and genius too.

THE HIPPOCAMPUS: ON TOP OF IT ALL

Three large brain structures lie under the neocortical sheet and communicate with it. They are the basal ganglia, the cerebellum, and the hippocampus. All three existed prior to the neocortex. With a very broad brush, we can say that the basal ganglia were the primitive motor system, the cerebellum learned precise timing relationships of events, and the hippocampus stored memories of specific events and places. To some extent, the neocortex has subsumed their original functions. For example, a human born without much of a cerebellum will have deficits in timing and will have to apply more conscious effort when moving but otherwise will be pretty normal.

We know that the neocortex is responsible for all complex motor sequences and can directly control your limbs. It isn't that the basal ganglia are unimportant, just that the neocortex has taken over a large part of motor control. Because of this, I have described the overall function of the neocortex independently of the basal ganglia and the cerebellum. Some scientists may disagree with this assumption, but it is one I have used in this book and in my work.

The hippocampus, however, is a different beast. It is one of the most heavily studied areas in the brain because it is essential to the formation of new memories. If you lose both halves of the hippocampus (like many parts of the nervous system, it exists in both the left and the right sides of the brain), you lose the ability to form most new memories. Without the hippocampus you can still talk, walk, see, and hear, and for brief periods of time appear almost normal. But in fact you are profoundly impaired: you can't remember anything new. You can remember a friend you knew before you lost the hippocampus, but you cannot remember a new person. Even if you met with your doctor five times a day for a year, each time would be like the first time. You would have no memory of events that occurred after the loss of your hippocampus.

For many years I hated to think about the hippocampus because it didn't make sense to me. It clearly is essential for learning, yet it isn't the ultimate repository for most of what we know. The neocortex is. The classic view of the hippocampus is that new memories are formed there, and later, over a period of days, weeks, or months, these new memories are transferred to the neocortex. This made no sense to me. We know that sights, sound, touch—our sensory data stream—flow directly into the sensory areas of the cortex without first passing through the hippocampus. It seemed to me that this sensory information should automatically form new memories in the cortex. Why do we need a hippocampus to learn? How could a separate structure such as the hippocampus interfere and prevent learning in the cortex, only later to transfer the information back to the cortex?

I decided to put the hippocampus aside, figuring that some

day its role would become clear. That day happened in late 2002, right around the time I began writing this book. One of my colleagues at the Redwood Neuroscience Institute, Bruno Olshausen, pointed out that the connections between the hippocampus and the neocortex suggest that the hippocampus is the top region of the neocortex, not a separate structure. In this view, the hippocampus occupies the peak of the neocortical pyramid, the top block in figure 5. The neocortex appeared on the evolutionary scene sandwiched between the hippocampus and the rest of the brain. Apparently this view of the hippocampus as being at the top of the cortical hierarchy was known for some time; I was just not aware of it. I had spoken to several hippocampus experts and asked them to explain how this seahorse-shaped structure could transfer memories to the cortex. No one could. And no one mentioned that the hippocampus was at the top of the cortical pyramid, probably because the hippocampus not only sits at the top of the cortical pyramid, but it still connects directly to many older parts of the brain.

Yet I instantly saw this new perspective as the solution to my confusion.

Think about information flowing from your eyes, ears, and skin into the neocortex. Each region of the neocortex tries to understand what this information means. Each region tries to understand the input in terms of the sequences it knows. If it does understand the input, it says, "I understand this, it is just part of the object I am already seeing. I won't pass on the details." If a region doesn't understand the current input, it passes it up the hierarchy until some higher region does. However, a pattern that is truly novel will escalate further and further up the hierarchy. Each successively higher region says, "I don't know what this is, I didn't anticipate it, why don't you higher-ups look at it?" The net effect is that when you get to the top of the cortical pyramid, what you have left is information that can't be understood by previous experience. You are left with the part of the input that is truly new and unexpected.

In a typical day we encounter many new things that make it to the top—for example, a story in the newspaper, the name of the person you met this morning, and the car accident you saw on the way home. It is these unexplained and unanticipated remainders, the new stuff, that enter the hippocampus and are stored there. This information won't be stored forever. Either it will be transferred down into the cortex below or it will eventually be lost.

I have noticed that, as I get older, I have trouble remembering new things. For example, my children remember the details of most of the theatrical plays they have seen in the last year. I can't. Perhaps it is because I have seen so many plays in my life that rarely do I see anything truly new. New plays fit into memories of past plays, and the information just doesn't make it to my hippocampus. For my children, each play is more novel and does reach the hippocampus. If this is true, we could say the more you know, the less you remember.

Unlike the neocortex, the hippocampus has a heterogeneous structure with several specialized regions. It's good at the unique task of quickly storing whatever patterns it sees. The hippocampus is in the perfect position, at the top of the cortical pyramid, to remember what is novel. It is also in the perfect position to recall these novel memories, allowing them to be stored in the cortical hierarchy, which is a somewhat slow process. You can instantly remember a novel event in the hippocampus, but you will permanently remember something in the cortex only if you experience it over and over, either in reality or by thinking of it.

AN ALTERNATE PATH UP THE HIERARCHY

Your cortex has a second major pathway for passing information from region to region, up the hierarchy. This alternate path starts with cells in layer 5, which project to the thalamus (a different part of the thalamus from the one we discussed earlier) and then from the thalamus up to the next higher region of cortex. Whenever two regions of cortex directly connect to each other in a hierarchical fashion, they also connect indirectly through the thalamus. This second pathway passes information only up the hierarchy, not down. So as we move up the cortical hierarchy, there is a direct path between two regions and an indirect path through the thalamus.

The second path has two modes of operation determined by cells in the thalamus. In one mode, the path is mostly shut down so information does not flow through. In the other mode, information flows accurately between regions. Two scientists, Murray Sherman of the State University of New York at Stony Brook and Ray Guillery of the University of Wisconsin School of Medicine, have described this alternate pathway and postulated that it might be as important as the direct one (perhaps more so), which has been the subject of this chapter so far. I have a speculation of what this second pathway is doing.

Read this word: *imagination.* Most people can recognize the word in a single glance, one fixation. Now look at the letter *i* in the middle of the word. Now look at the dot over that *i.* Your eyes can be fixated on the exact same location, but in one case you see the word, in the next case you see the letter, and in the final case you see the dot. Stare at the *i* and try to alternate your perception between the word, the letter, and the dot. If you have trouble, try saying "dot," "i," and "imagination" while you stare at the dot. In all cases the exact same information is entering V1, yet by the time it gets to some higher region like IT you perceive different things, different levels of detail. Region IT knows how to recognize all three objects. It can recognize the dot in isolation, the letter *i*, and the entire word at a glance. But when you perceive the entire word, V4, V2, and V1 handle the details, and all IT gets to know about is the word. You normally don't perceive the individual letters while reading; you perceive words or phrases. But you can perceive the letters if you choose to. We are doing this sort of attentional shift all the time, but we aren't generally aware of it. I can be listening to music playing in the background and barely be aware of the melody, but if I try, I can isolate the singer or the bass guitar. The same sound is entering my head, but I can focus my perceptions. Every time you scratch your head, the movement makes a loud internal sound, but usually you are unaware of the noise. If you focus on it, though, you can hear the sound clearly. This is another example of sensory input that normally is handled low in the cortical hierarchy but can be brought to higher levels if you attend to it.

I speculate that the alternate pathway through the thalamus is the mechanism by which we attend to details that normally we wouldn't notice. It bypasses the grouping of sequences in layer 2, sending the raw data to the next higher region of cortex. Biologists have shown that the alternate pathway can be turned on in one of two ways. One is by a signal from the higher region of the cortex itself. This is the method you used when I asked you to attend to details that normally you wouldn't notice, such as the dot on the *i* or the sound of a head scratch. The second method that can activate this pathway is a large, unexpected signal from below. If the input to the alternate pathway is strong enough, it sends a wake-up signal to the higher region, which again can turn on the pathway. For example, if I showed you a face and asked you what it was, you would say "Face." If I showed you the same face but it had a strange mark on the nose, you would first recognize the face, but then immediately your lower levels of vision would notice that something is wrong. This error forces the opening of the attentional pathway. The details will now take the alternate path, bypassing the grouping that normally occurs, and your attention will be drawn to the mark. You now see the mark and not just the face. If it was sufficiently odd, the mark could occupy your entire attention. In this way, unusual events quickly rise to your attention. This is why we can't avoid focusing on deformities and other unusual patterns. Your brain does this automatically. Often, however, errors aren't strong enough to open the alternate pathway. This is why we sometimes don't notice if a word is misspelled as we read it.

CLOSING THOUGHTS

To find and establish a new scientific framework, it is necessary to look for the simplest concepts capable of uniting and explaining what were large quantities of disparate facts. It is an unavoidable consequence of this process that the pendulum swings too far toward simplification. Important details are likely to be ignored, and facts will be misinterpreted. If the framework takes hold, refinements and fixes will inevitably be found showing where the initial proposal went too far, didn't go far enough, or was in error.

In this chapter, I have introduced many speculative ideas on how the neocortex works. I expect that several of these ideas will prove to be wrong, and probably all of these ideas will be revised. There are also many details I haven't even mentioned. The brain is very complex; neuroscientists reading this book will know that I have presented a crude characterization of the complexities of the real brain. Yet I believe the framework as a whole to be sound. All I can hope is that the core ideas will be preserved as the details change in the face of new data and understanding.

Finally, you may be struggling with the idea that a simple but large memory system could actually result in all that humans do. Could you and I really be just a hierarchical memory system? Could our lives, beliefs, and ambitions be stored in trillions of tiny synapses? In 1984, I started writing computer programs professionally. I had written small programs before but this was my first time programming a computer with a graphical user interface, and my first time working on large and sophisticated applications. I was writing software for an operating system created by Grid Systems. With windows, multiple fonts, and menus, the Grid operating system was quite advanced for its time.

One day I was struck by the near impossibility of what I was doing. As a programmer, I wrote one line of code at a time. I grouped lines of code into blocks called subroutines. Subroutines were grouped into modules. Modules were combined to form an application. The spreadsheet program I was working on had so many subroutines and modules that no one person could understand the entire thing. It was complex. Yet a single line of code would do very little. To place one pixel on the display took several lines of code. To draw an entire screen for the spreadsheet required the computer to execute millions of instructions that were spread over hundreds of subroutines. Subroutines invoked other subroutines in repetitive and recursive ways. It was so complicated that it wasn't possible to know everything that would happen when the program ran. It struck me as highly improbable that when the program did run it would draw its image in what seemed like an instant. Its outward appearance was tables of numbers, labels, text, and graphs. It behaved like a spreadsheet. But I knew what was going on inside the computer with the processor executing one simple instruction at a time. It was hard to believe that the computer could find its way through the maze of modules and subroutines and execute all those instructions so quickly. If I hadn't known better, I would have been certain the whole thing couldn't work. I realized that if someone had invented the concept of a computer with a graphical user interface and a spreadsheet application, and presented it to me on paper, I would have rejected it as impractical. I would have said it would take forever to do anything. It was a humbling thought because it did work. It was then that I realized my intuitive sense for the speed of the microprocessor and my intuitive sense for the power of hierarchical design were inadequate.

There is a lesson here about the neocortex. It isn't made of superfast components and the rules under which it operates are not that complex. However, it does have a hierarchical structure that contains billions of neurons and trillions of synapses. If we find it hard to imagine how such a logically simple but numerically vast memory system can create our consciousness, our languages, our cultures, our art, this book, and our science and technology, I suggest it is because our intuitive sense of the capacity of the cortex and the power of its hierarchical structure is inadequate. The neocortex does work. It isn't magic. We can understand it. And like a computer, ultimately we can build intelligent machines that work on the same principles.

7

CONSCIOUSNESS AND CREATIVITY

When I give talks about my brain theory, audiences are usually quick to grasp the significance of prediction as it relates to a host of human activities. They ask many related questions. Where does creativity come from? What is consciousness? What is imagination? How can we separate reality from false beliefs? Although these topics have not been in the forefront of my motivations for studying brains, they are of interest to nearly everyone. I don't pretend to be an expert in these topics, but the memory-prediction framework of intelligence can provide some answers and useful insights. In this chapter, I address some of the most frequently asked questions.

ARE ANIMALS INTELLIGENT?

Is a rat intelligent? Is a cat intelligent? When did intelligence begin in evolutionary time? I love this question because I find the answer surprising.

Everything I have written so far about the neocortex and how it works depends on a very basic premise—that the world has structure and is therefore predictable. There are patterns in the world: faces have eyes, eyes have pupils, fires are hot, gravity makes objects fall, doors open and shut, and so forth. The world is not random, nor is it homogeneous. Memory, prediction, and behavior would be meaningless if the world was without structure. All behavior, whether it is the behavior of a human, a snail, a single-cell organism, or a tree, is a means of exploiting the structure of the world for the benefit of reproduction.

Imagine a one-cell animal living in a pond. The cell has a flagellum that lets it swim. On the surface of the cell are molecules that detect the presence of nutrients. Since not all areas of the pond have the same concentration of nutrients, there is a gradual change in value, or gradient, of nutrients from one side of the cell to the other. As it swims across the pond, the cell can detect the shift. This is a simple form of structure in the world of the one-cell animal. The cell exploits its chemical awareness by swimming toward places with higher concentrations of nutrients. We could say that this simple organism is making a prediction. It is predicting that by swimming in a certain way it will find more nutrients. Is there memory involved in this prediction? Yes, there is. The memory is in the DNA of the organism. The one-cell animal did not learn, in its lifetime, how to exploit this gradient. Rather, the learning occurred over evolutionary time and is stored in the animal's DNA. If the structure of the world changed suddenly, this particular one-cell animal could not learn to adapt. It could not alter its DNA or the resulting behavior. For this species, learning can occur only through evolutionary processes over many generations.

Is this one-cell organism intelligent? Using the everyday notion of human intelligence, the answer is no. But the animal does lie at the far edge of a continuum of species that use memory and prediction to reproduce more successfully, and by that more academic measure the answer is yes. The point is not to label some species as intelligent and others as not intelligent. Memory and prediction are used by all living things. There is just a continuum of methods and sophistication in how they do it.

Plants also use memory and prediction to exploit the structure of the world. A tree makes a prediction when it sends its roots down into the soil and its branches and leaves up toward the sky. The tree is predicting where it will find water and minerals based on the experience of its ancestors. Of course a tree doesn't think; its behavior is automatic. But the species is exploiting the structure of the world in the same way as the one-cell organism. Every plant species has a distinct set of behaviors that exploit slightly different parts of the structure of the world.

Eventually, plants evolved communication systems, based mostly on the slow release of chemical signals. If an insect damages part of a tree, the tree sends chemicals through its vascular system to other parts of the tree, which triggers a defense system, such as making toxins. Through such a communication system, the tree can exhibit slightly more complex behavior. Neurons probably evolved as a way to communicate information more quickly than a plant's vascular system. You could think of a neuron as just a cell with its own vascular appendages. At some point, instead of slowly moving chemicals along these appendages, the neuron started using electrochemical spikes, which travel much faster. In the beginning, fast synaptic transmission and simple nervous systems probably did not involve much if any learning. The name of the game was simply faster signaling.

But then, in the march of evolutionary time, something really interesting happened. Connections between neurons became modifiable. A neuron could send a signal or not send a signal, depending on what had happened recently. Behavior could now be modified within the life of an organism. The nervous system became plastic, and so did behavior. Because memories could be rapidly formed, the animal could learn the structure of its world during its own lifetime. If the world suddenly changed—say, a new predator arrived on the scene—the animal didn't have to stick with its genetically determined behavior, which might no longer be appropriate. Plastic nervous systems became a tremendous evolutionary advantage and led to a burst of new species from fish to snails to humans.

As we saw in chapter 3, all mammals have an old brain, on top of which sits the neocortex. The neocortex is just the most recent neural tissue to evolve. But with its hierarchical structure, invariant representations, and prediction by analogy, the cortex allows mammals to exploit much more of the structure of the world than an animal without a neocortex can. Our cortically endowed ancestors could envision how to make a net and catch fish. The fish are not able to learn that nets mean death or to figure out how to build tools to cut nets. All mammals, from rats to cats to humans, have a neocortex. They are all intelligent, but to differing degrees.

WHAT'S DIFFERENT ABOUT HUMAN INTELLIGENCE?

The memory-prediction framework offers two answers to this question. The first is pretty straightforward: our neocortex is larger than, say, a monkey's or a dog's. By enlarging the cortical sheet to the size of a large dinner napkin, our brains can learn a more complex model of the world and make more complex predictions. We see deeper analogies, more structure on structure, than other mammals. If we want to find a mate we don't just look at simple attributes such as health, we interview their friends and parents, we observe how they drive and speak, and judge how honest they are. We look at these secondary and tertiary attributes to try to predict how our potential mate will behave in the future. Stock market traders look for structure in trading patterns. Mathematicians look for structure in numbers and equations. Astronomers look for structure in the motions of the planets and the stars. Our larger neocortex allows us to see our home as part of a town, which is part of a region, which is part of a planet, which is part of a large universe—structure within structure. No other mammal can ruminate to this depth. I am pretty certain my cat has no concept of a world outside our house.

The second difference between the intelligence of humans and other mammals is that we have language. Entire books have been written on the supposedly unique properties of language and how it developed. However, language fits nicely into the memory-prediction framework without any special language sauce or dedicated language machinery. Spoken and written words are just patterns in the world, as are melodies, cars, and houses. The syntax and semantics of language are not different from the hierarchical structure of other everyday objects. And in the same way that we associate the sound of a train with the visual memory image of a train, we associate spoken words with our memory of their physical and semantic counterparts. Through language one human can invoke memories and create new juxtapositions of mental objects in another human. Language is pure analogy, and through it we can cause other humans to experience and learn about things they may never actually see. The development of language required a large neocortex capable of handling the nested structure of syntax and semantics. It also required a more fully developed motor cortex and musculature to enable us to make sophisticated, highly articulate sounds or gestures. With language, we can take patterns that we learn in a lifetime and transmit them to our children and our tribe. Language, whether it be written, spoken, or embodied in cultural traditions, became the means by which we pass on what we know about the world from generation to generation. Today, printed and electronic communications allow us to share our knowledge with millions of people around the world. Animals without language don't transmit nearly as much information to their offspring. A rat can learn many patterns in its lifetime, but it doesn't pass on detailed new information—"Hey junior, here's how my father taught me to avoid electric shocks."

Thus, intelligence could be traced over three epochs, each using memory and prediction. The first would be when species used DNA as the medium for memory. Individuals could not learn and adapt within their lifetimes. They could only pass on the DNA-based memory of the world to their offspring through their genes.

The second epoch began when nature invented modifiable nervous systems that could quickly form memories. An individual could now learn about the structure of its world and adapt its behavior accordingly within its lifetime. But an individual still could not communicate this knowledge to its offspring other than by direct observation. The creation and expansion of the neocortex occurred within this second epoch, but did not define it.

The third and final epoch is unique to humans. It begins with the invention of language and the expansion of our large neocortex. We humans can learn a lot of the structure of the world within our lifetimes, and we can effectively communicate this to many other humans via language. You and I are participating in this process right now. I have spent a large part of my life searching for the structure in brains and how that structure leads to thought and intelligence. Through this book I am spreading what I have learned to you. Of course I couldn't have done this if I hadn't had access to the knowledge gathered by hundreds of scientists, who learned from others, and so on down through the ages. I was able to assimilate and add to what others have written about their own thinking and observation.

We have become the most adaptable creatures on the planet and the only ones with the ability to transfer our knowledge of the world broadly within our populace. The human population has undergone explosive growth because we can learn and exploit so much of the structure of the world and communicate it to other humans. We can thrive anywhere, be it a rain forest, a desert, the frozen tundra, or the concrete jungle. The combination of a large neocortex and language has led to the spiraling success of our species.

WHAT IS CREATIVITY?

I am frequently asked about creativity, I suspect because many people see creativity as something a machine couldn't do, and therefore it is a challenge to the entire idea of building intelligent machines. What is creativity? We have already encountered the answer several times in this book. Creativity is not something that occurs in a particular region of the cortex. Nor is it like emotions or balance, which are rooted in particular structures and circuits outside of the cortex. Rather, creativity is an inherent property of every cortical region. It is a necessary component of prediction.

How can this be true? Isn't creativity some extraordinary quality that requires high intelligence and giftedness? Not really. Creativity can be defined simply as making predictions by analogy, something that occurs everywhere in cortex and something you do continually while awake. Creativity occurs along a continuum. It ranges from simple everyday acts of perception occurring in sensory regions of the cortex (hearing a song in a new key) to difficult, rare acts of genius occurring at the highest levels in the cortex (composing a symphony in a brand-new way). At a fundamental level, everyday acts of perception are similar to the rare flights of brilliance. It is just that the everyday acts are so common we don't notice them.

By now you have a basic understanding of how we create invariant memories, how we use invariant memories to make predictions, and how we make predictions of future events that are always somewhat different from anything we have experienced in the past. Recall also that our invariant memories are of sequences of events. We make predictions by combining the invariant memory recall of what should happen next with the details pertaining to this moment in time (remember the parable of predicting when the train will arrive). Prediction is the application of invariant memory sequences to new situations. Therefore all cortical predictions are predictions by analogy. We predict the future by analogy to the past.

Imagine you are about to have dinner in an unfamiliar restaurant and you want to wash your hands. Even though you have never been in this building before, your brain predicts that there will be a restroom somewhere in the restaurant with a basin suitable for hand washing. How does it know this? Other restaurants you have been in have a restroom, and by analogy this restaurant will likely have one, too. Further, you know where and what to look for. You predict there will be a door or sign with some type of symbol associated with men or women. You predict it will be toward the back of the restaurant, either by the bar or down a hall, but generally not in plain view of the eating areas. Again, you have never been in this particular restaurant before, but by analogy to other eating establishments you are able to find what you need. You don't look around randomly. You look for expected patterns that let you find the restroom quickly. This kind of behavior is a creative act; it is predicting the future by analogy to the past. We don't normally think of this as being creative, but it very much is.

Recently I bought a vibraphone. We have a piano, but I had never played the vibraphone before. The day we brought it home, I took a sheet of music from the piano, placed it on the stand over the vibraphone, and started playing simple melodies. My ability to do this was not remarkable. But in a fundamental way, it was a creative act. Think about what was involved. I have an instrument that is very different from a piano. The vibraphone has gold metal bars; the piano has black and white keys. The gold bars are big and gradually change in size; the keys are small and of two different sizes. The gold bars are arranged in two different rows; the black and white keys are interleaved. On one instrument I use my fingers, and on the other I swing mallets. For this I'm standing up, and for that I'm sitting down. The particular muscles and motions needed to play the vibraphone are completely different from those needed to play the piano.

So how was I able to play a melody on an unfamiliar instrument? The answer is that my cortex sees an analogy between the keys on a piano and the bars on a vibraphone. Using this similarity allowed me to play a tune. It isn't really any different from singing a song in a new key. In both cases, we know what to do by analogy to past learning. I realize that to you the similarity between these two instruments may appear obvious, but that is only because our brains automatically see analogies. Try to program a computer to find similarities between objects such as pianos and vibraphones and you will see how incredibly difficult this is. Prediction by analogy—creativity—is so pervasive we normally don't notice it.

We do, however, believe we are being creative when our memory-prediction system operates at a higher level of abstraction, when it makes uncommon predictions, using uncommon analogies. For example, most people would agree that a mathematician who proves a difficult conjecture is being creative. But let's take a close look at what's involved with her mental efforts. Our mathematician stares hard at an equation and says, "How am I going to tackle this problem?" If the answer isn't readily obvious she may rearrange the equation. By writing it down in a different fashion, she can look at the same problem from a different perspective. She stares some more. Suddenly she sees a part of the equation that looks familiar. She thinks, "Oh, I recognize this. There's a structure to this equation that is similar to the structure of another equation I worked on several years ago." She then makes a prediction by analogy. "Maybe I can solve this new equation using the same techniques I used successfully on the old equation." She is able to solve the problem by analogy to a previously learned problem. It is a creative act.

My father had a mysterious blood disorder that his physicians could not diagnose. So how did they know what treatment to offer? One of the things they did was to look at months of data taken from analyses of my father's blood to see if they could identify patterns. (My father printed a beautiful chart so the doctors could see the data clearly.) While his symptoms did not closely match those of known diseases, there were some similarities. The doctors ended up basing his treatment on a mixture of strategies that had worked for other blood disorders. The treatments used were guesses based on analogies to diseases the physicians had previously treated. Recognizing these patterns required extensive exposure to other uncommon diseases.

Shakespeare's metaphors are the paragon of creativity. "Love is a smoke made with the fume of sighs." "Adversity's sweet milk, philosophy." "There's daggers in men's smiles." Such metaphors become obvious when you see them but they're very hard to invent, which is one reason why Shakespeare is regarded as a literary genius. To create such metaphors he had to see a succession of clever analogies. When he writes "There's daggers in men's smiles," he is not talking about daggers or smiles. Daggers are analogous to ill intent, and men's smiles are analogous to deceit. Two clever analogies in only five words! At least that is how I interpret it. Poets have the gift of correlating seemingly unrelated words or concepts in manners that illuminate the world in new ways. They create unexpected analogies as a means of teaching higher-level structure.

In fact, highly creative works of art are appreciated because they violate our predictions. When you see a film that breaks the familiar mold of a character, story line, or cinematography (including special effects), you like it because it is not the same old same old. Paintings, music, poetry, novels—all creative artistic forms—strive to break convention and violate the expectations of an audience. There is a contradictory tension in what makes a work of art great. We want art to be familiar yet at the same time to be unique and unexpected. Too much familiarity is retread or kitsch; too much uniqueness is jarring and difficult to appreciate. The best works break some expected patterns while simultaneously teaching us new ones. Consider a great piece of classical music. The best music has an appeal at a simple level—good beat, simple melody and phrasing. Anyone can understand and appreciate it. However, it is also a little different and unexpected. But the more you listen to it, the more you see there is pattern in the unexpected parts, such as repeated unusual harmonies or key changes. The same is true with great literature or great movies. The more you read or see them, the more creative detail and complexity of structure you observe.

You've probably had the experience of looking at something when a twinge of recognition goes off in your head: "Hmmm, I've seen this pattern before, someplace else . . ." You may not have been trying to solve a problem, it's just that an invariant representation in your brain was activated by a novel situation. You saw an analogy between two normally unrelated events. I might recognize that promoting a scientific idea is similar to selling a business idea or that bringing about political reform is like raising children. If I'm a poet, *voilà!*, I have a new metaphor. If I'm a scientist or engineer, I have a new solution to a long-standing problem. Creativity is mixing and matching patterns of everything you've ever experienced or come to know in your lifetime. It's saying "this is kinda like that." The neural mechanism for doing this is everywhere in the cortex.

ARE SOME PEOPLE MORE CREATIVE THAN OTHERS?

A related question I often hear is, "If all brains are inherently creative, why are there differences in our creativity?" The memory-prediction framework points to two possible answers. One has to do with nature and the other with nurture.

On the nurture side, everyone has different life experiences.

Therefore everyone develops different models and memories of the world in his or her cortex, and will make different analogies and predictions. If I have been exposed to music, I will be able to sing songs in new keys and play simple melodies on new instruments. If I have never been exposed to music, I will not be able to make these predictive leaps. If I have studied physics, I will be able to explain the behavior of everyday objects via analogy to the laws of physics. If I grew up with dogs, I am apt to see analogies about dogs and will be better at predicting their behavior. Some people are more creative in social situations or in language, math, or diplomacy, all based on the environment they grew up in. Our predictions, and thus our talents, are built upon our experiences.

In chapter 6, I described how memories are pushed down the cortical hierarchy. The more you are exposed to certain patterns, the more the memory of these patterns are re-formed at lower levels. This allows you to learn the relationships among higher-order abstract objects at the top. It's the essence of expertise. An expert is someone who through practice and repeated exposure can recognize patterns that are more subtle than can be recognized by a nonexpert, such as the shape of a fin on a late-fifties car or the size of a spot on a seagull's beak. Experts can recognize patterns on top of patterns. Ultimately there is a physical limit to what we can learn constrained by the size of our cortex. But as humans, our cortex is large compared to other species and we have a tremendous flexibility in what we can learn. It all depends on what we are exposed to throughout our lives.

On the nature side, brains exhibit physical variation. Certainly some of the differences are genetically determined such as the size of regions (individuals can show as much as a three-fold difference in the gross area of V1) and hemispheric laterality (women tend to have thicker cables connecting the left and right sides of the brain than men do). Among humans, some brains probably have more cells or different kinds of connections. It's unlikely that Albert Einstein's creative genius was purely a function of the stimulating environment in the patent office where he worked as a young man. Recent analyses of his brain—which had been thought lost, but was found preserved in a jar a few years ago—reveal that his brain was measurably unusual. It had more support cells, called glia, per neuron than average. It showed an unusual pattern of grooves, or sulci, in the parietal lobes—a region thought to be important for mathematical abilities and spatial reasoning. It was also 15 percent wider than most other brains. We may never know why Einstein was as creative and smart as he was, but it is a safe bet that part of his talent derived from genetic factors.

Whatever the difference between brilliant and average brains, we are all creative. And through practice and study we can enhance our skills and talents.

CAN YOU TRAIN YOURSELF TO BE MORE CREATIVE?

Yes, most definitely. I have found there are ways to foster finding useful analogies when working on problems. First, you need to assume up front that there is an answer to what you are trying to solve. People give up too easily. You need confidence that a solution is waiting to be discovered and you must persist in thinking about the problem for an extended period of time.

Second, you need to let your mind wander. You need to give your brain the time and space to discover the solution. Finding a solution to a problem is literally finding a pattern in the world, or a stored pattern in your cortex that is analogous to the problem you are working on. If you are stuck on a problem, the memory-prediction model suggests that you should find different ways to look at it to increase the likelihood of seeing an analogy with a past experience. If you just sit there and stare at it over and over, you won't get very far. Try taking the parts of your problem and rearranging them in different ways—literally and figuratively. When I play Scrabble, I constantly shuffle the order of the tiles. It isn't that I hope the letters will by chance spell a new word, but that different letter combinations will remind me of words or parts of words that might be part of a solution. If you are looking at a drawing of something that just doesn't make sense, try drawing it upside down, changing colors, or changing perspectives. For example, when I was thinking about how different patterns in V1 could lead to invariant representations in IT, I was stuck. So I flipped the problem around and asked how a constant pattern in IT could lead to different predictions in V1. Inverting the problem was immediately helpful, ultimately leading to my belief that V1 should not be viewed as a single cortical region.

If you get stuck on a problem, go away for a little while. Do something else. Then start again, rephrasing the problem anew. If you do this enough times something will click sooner or later. It may take days or weeks, but eventually it will happen. The goal is to find an analogous situation somewhere in your past or present experience. To succeed you must ponder the problem often but also do other things so the cortex will have the opportunity to find an analogous memory.

Here is another example of how rearranging a problem led to a novel solution. In 1994, my colleagues and I were trying to figure out how to enter text on handheld computers. Everyone was focused on handwriting recognition software. They said, "Look, you write things on pieces of paper, you should be able to write the same way on a computer screen." Unfortunately, this turns out to be really hard. It's another one of those things that computers aren't very good at, even though brains find it quite simple. The reason is that the brain uses memory and current context to predict what is written. Words and letters that are unrecognizable on their own are easily recognized in context. Pattern matching with computers is not sufficient to the task. I had designed several computers that used traditional handwriting recognition but it was never good enough.

I struggled with how to make the recognition software work better for several years and was stuck. One day I stepped back and decided to look at the problem from a different perspective. I looked for analogous problems. I said to myself, "How do we enter text into desktop computers? We type on a keyboard. How do we know how to type on a keyboard? Well, actually, it's not easy. It's a recent invention and it takes a long time to learn. Touch-typing on a typewriter-style keyboard is hard and not intuitive, it isn't at all like writing—yet millions of people learn how. Why? Because it works." My thinking continued by analogy, "Maybe I can come up with a text input system that is not necessarily intuitive, that you have to learn, but people will use it because it works."

Literally, that's the process I went through. I used the act of typing on a keyboard as an analogy to figure out how to enter text with a stylus on a display. I recognized that people were willing to learn a difficult task (typing) because it was a reliable and fast way to enter text into a machine. Therefore if we could create a new method of entering text with a stylus that was fast and reliable, people would use it even though it required learning. So I designed an alphabet that would reliably translate what you wrote into computer text; we called it Graffiti. With traditional handwriting recognition systems, when the computer misinterprets your writing you don't know why. But the Graffiti system always produces the correct letter unless you make a mistake in writing. Our brains hate unpredictability, which is why people hate traditional handwriting recognition systems.

Many people thought Graffiti was a sensationally stupid idea. It went against everything they believed about how computers were supposed to work. The mantra in those days was that computers should adapt to the user, not the other way around. But I was confident that people would accept this new way of entering text by analogy to the keyboard. Graffiti turned out to be a good solution and was widely adopted. To this day I still hear people claim that computers should adapt to users. This isn't always true. Our brains prefer systems that are consistent and predictable, and we like learning new skills.

CAN CREATIVITY LEAD ME ASTRAY? CAN I FOOL MYSELF?

False analogy is always a danger. The history of science is rife with examples of beautiful analogies that turned out to be wrong. For example, the celebrated astronomer Johannes Kepler convinced himself that the orbits of the six known planets were defined by the Platonic solids. The Platonic solids are the only three-dimensional shapes that can be constructed entirely out of regular polygons. There are exactly five of them: tetrahedron (four equilateral triangles), sextahedron (six squares, aka a cube), octahedron (eight equilateral triangles), dodecahedron (twelve regular pentagons), and icosahedron (twenty equilateral triangles). They were discovered by the ancient Greeks, who were obsessed with the relationship of mathematics and the cosmos.

Like all Renaissance scholars, Kepler was heavily influenced by Greek thought. It seemed to him that it couldn't possibly be a coincidence that there were five Platonic solids and six planets. As he put it in his book *The Cosmic Mystery* (1596): "The dynamic world is represented by the flat-faced solids. Of these there are five: when viewed as boundaries, however, these five determine six distinct things: hence the six planets that revolve about the sun. This is also the reason why there are but six planets." He saw a beautiful but entirely false analogy.

Kepler went on to account for the orbits of the planets in terms of nested Platonic solids that were all centered on the sun. He took the sphere defined by Mercury's orbit as his baseline and circumscribed it with an octahedron. The tips of the octahedron defined a larger sphere, which gave the orbit of Venus.

Around Venus's orbit he circumscribed an icosahedron whose outer tips yielded the orbit of the Earth. The progression continued: a dodecahedron drawn around the Earth's orbit gave the orbit of Mars, a tetrahedron around Mars's orbit gave the orbit of Jupiter, and a cube around Jupiter's orbit gave the orbit of Saturn. It was elegant and beautiful. Given the limited precision of astronomical data in his day, he was able to convince himself that this scheme worked! (Years later, Kepler realized he had been mistaken after he got hold of the high-precision astronomical data of his deceased colleague Tycho Brahe, which proved that planetary orbits are ellipses, not circles.)

Kepler's excitement serves as a cautionary tale for scientists, and indeed for all thinkers. The brain is an organ that builds models and makes creative predictions, but its models and predictions can as easily be specious as valid. Our brains are always looking at patterns and making analogies. If correct correlations cannot be found, the brain is more than happy to accept false ones. Pseudoscience, bigotry, faith, and intolerance are often rooted in false analogy.

WHAT IS CONSCIOUSNESS?

This is one of those questions neuroscientists dread, unnecessarily so in my opinion. Some scientists, such as Christof Koch, are willing to tackle the issue of consciousness, but most consider it a question of philosophy bordering on pseudoscience. I think it deserves consideration if for no other reason than many people are curious about it. I cannot provide a completely satisfactory answer, but I think memory and prediction can address part of it. First, here's a flavor of the conundrum as it comes up in conversation.

Not long ago I was at a science conference in a lovely spot on Long Island Sound. It was early evening when a dozen of us took our glasses of wine down to a pier to sit by the water and chat before dinner and the evening session. After a while, the conversation turned to the topic of consciousness. As I said, neuroscientists normally don't talk about this, but we were in a beautiful setting, some wine had been consumed, and the topic was brought up.

A British scientist was holding forth on her ideas about consciousness and said, "Of course, we'll never understand consciousness." I disagreed, "Consciousness is not a big problem. I think consciousness is simply what it feels like to have a cortex." A silence fell on the group, then an argument quickly ensued as several scientists tried to educate me on my obvious error. "You must admit that the world seems so alive and beautiful. How can you deny your consciousness that perceives the world? You must admit you feel like something special." To make a point, I said, "I don't know what you are talking about. Given the way you are talking about consciousness, I have to conclude I am different from you. I don't feel what you are feeling, so maybe I am not a conscious being. I must be a zombie." Zombies are often invoked when philosophers talk about consciousness. A zombie is defined as physically identical to a human, but lacking consciousness. They are walking breathing meat machines, but with no one home.

The British scientist looked at me. "Of course you are conscious."

"No, I don't think so. I may look that way to you, but I'm not a conscious human being. Don't worry about it, I'm okay with it."

She said, "Well, don't you perceive the wonder?" and swept her arm toward the glistening water as the sun began to sink and the sky turned iridescent salmon-pink.

"Yes, I see all this stuff. So?"

"Then how do you explain your subjective experience?"

I replied, "Yes, I know I'm here. I have memories of things like this evening. But I don't feel anything special is going on, so if you feel something special maybe I'm just not conscious." I was trying to pin her down about what she thought was so miraculous and unexplainable about consciousness. I was trying to get her to define consciousness.

We continued this line of argument—a yes you are, no I'm not kind of thing—until it was time to head up to dinner. I don't think I changed anyone's mind about the existence and meaning of consciousness. But I was trying to get them to realize that most people think consciousness is some kind of magical sauce that is added on top of the physical brain. You've got a brain, made of cells, and you pour consciousness, this magical sauce, on it, and that's the human condition. In this view, consciousness is a mysterious entity separate from brains. That's why zombies have brains but they don't have consciousness. They have all the mechanical stuff, neurons and synapses, but they don't have the special sauce. They can do everything a human can do. From the outside you can't tell a zombie from a human.

The idea that consciousness is something extra stems from earlier beliefs in élan vital—a special force once thought to animate living things. People believed you needed this life force to explain the difference between rocks and plants or metals and maidens. Few people believe this anymore. Nowadays we know enough about the differences between inanimate and animate matter to understand that there isn't a special sauce. We now know a great deal about DNA, protein folding, gene transcription, and metabolism. While we don't yet know all the mechanisms of living systems, we know enough about biology to leave out magic. Similarly, no longer do people suggest it takes magic or spirits to make muscles move. We have folding proteins that pull long molecules past one another. You can read all about it.

Nevertheless, many people persist in believing that consciousness is different and can't be explained in reductionist biological terms. Again, I am not a student of consciousness. I haven't read all the philosophers' opinions. But I have some ideas about what I think people are confusing in this debate. I believe consciousness is simply what it feels like to have a neocortex. But we can do better than that. We can break consciousness into two major categories. One is similar to self-awareness—the everyday notion of being conscious. This is relatively easy to understand. The second is *qualia*—the idea that feelings associated with sensation are somehow independent of sensory input. Qualia is the harder part.

When most people say the word *conscious,* they are referring to the first category. "Were you conscious that you walked past me without saying hello?" "Were you conscious when you fell out of bed last night?" "You aren't conscious when you sleep." Some people say this form of consciousness is exactly the same as awareness. The two are close, but I don't think awareness quite captures it correctly. I suggest this meaning of consciousness is synonymous with forming declarative memories. Declarative memories are memories that you can recall and talk about to someone else. You can express them verbally. If you ask me where I went last weekend, I can tell you. That is a declarative memory. If you ask me how to balance a bicycle, I can tell you to hold the handle bar and push the pedals, but I can't explain exactly how to do it. How to balance a bicycle has mostly to do with neural activity in the old brain, so it is not a declarative memory.

I have a little thought experiment to show how our everyday notion of consciousness is the same as forming declarative memories. Recall that all memory is believed to reside in physical changes to synapses and the neurons they connect to. Therefore, if I had a method to reverse those physical changes, your memory would be erased. Now imagine I could flip a switch and return your brain to the exact physical state it was in at some point in the past. It could be an hour ago, twenty-four hours ago, whatever. I just flip the switch in my way-back machine and your synapses and neurons return to a previous state in time. By doing so, I erase all your memory of what occurred since that time.

Let's assume you go through today and wake up tomorrow. But just as you're waking up, I flip the switch and erase the last twenty-four hours. You would have absolutely zero memory of the previous day. From your brain's perspective, yesterday never happened. I would tell you it's Wednesday and you'd protest, "No, it's Tuesday. I'm certain of it. The calendar has been altered. No way, this is Tuesday. Why are you pulling this trick on me?" But everyone whom you had met on Tuesday would say that you had been conscious throughout the day. They saw you, had lunch with you, and talked with you. Don't you remember it? You'd say no, it didn't happen. Finally, shown a video of you having lunch, you gradually become convinced that the day did happen, even though you have no memory of it. It's as if you were a zombie for a day, not conscious. However, you were conscious at the time. Your belief that you were conscious disappeared only when your declarative memory was erased.

This thought experiment captures the equivalence between declarative memory and our everyday notion of being conscious. If during and at the end of a game of tennis I ask you if you are conscious, you would, of course, say yes. If I then erased your memory of the last two hours, you would claim to have been unconscious and not responsible for your actions during that time. In either case, you played the same game of tennis. The only difference is whether you have a memory of it at the time I ask you. Therefore, this meaning of consciousness is not absolute. It can be changed after the fact by memory erasure.

The more difficult question about consciousness concerns qualia. Qualia is often phrased in Zen-like queries, such as "Why is red red and green green? Does red look the same to me as it does to you? Why is red emotionally laden with certain feelings? It has a certain inextricable quality or feelingness to me. What feelingness does it cause in you?"

I find such descriptions difficult to relate to neurobiology, so I'd like to rephrase the question. For me, an equivalent question, but one I still find hard to explain, is, Why do different senses seem qualitatively different? Why does sight seem different from hearing and why does hearing seem different from touch? If the cortex is the same everywhere, if it works with the same processes, if it is just dealing with patterns, if no sound or light enters the brain, just patterns, then why does vision seem so different from hearing? I find it difficult to describe how sight differs from hearing, but it self-evidently is. I assume it is for you, too. Yet an axon representing sound and another representing light are, for all practical purposes, identical. "Lightness" and "soundness" are not carried down the axon of a sensory neuron.

People with a condition called synesthesia have brains that blur the distinction between the senses—certain sounds have a color, or certain textures have a color. This tells us that the qualitative aspect of a sense is not immutable. Through some sort of physical modification, a brain can impart a qualitative aspect of vision to an auditory input.

So what is the explanation for qualia? I can think of two possibilities, neither of which I find completely satisfactory. One is that although hearing, touch, and vision work under similar principles in the neocortex, they are handled differently below the cortex. Hearing relies on a set of audition-specific subcortical structures that process auditory patterns before they reach the cortex. Somatosensory patterns also travel through a set of subcortical areas that are unique to somatic senses. Perhaps qualia, like emotions, are not mediated purely by the neocortex. If they are somehow bound up with subcortical parts of the brain that have unique wiring, perhaps tied to emotion centers, this might explain why we perceive them differently, even if it doesn't help explain why there is any sort of qualia sensation in the first place.

The other possibility I can think of is that the structure of the inputs—differences in the patterns themselves—dictates how you experience qualitative aspects of the information. The nature of the spatial-temporal pattern on the auditory nerve is different from the nature of the spatial-temporal pattern on the optic nerve. The optic nerve has a million fibers and carries quite a bit of spatial information. The auditory nerve has only thirty thousand fibers and carries more temporal information. These differences may be related to what we call qualia.

We can be certain that however consciousness is defined, memory and prediction play crucial roles in creating it.

Related to consciousness are the notions of mind and soul.

As a child I used to wonder what it would have been like if "I" had been born in another child's body in another country, as if "I" was somehow independent of my body. These feelings of a mind independent of physicalness are common and a natural consequence of how the neocortex works. Your cortex creates a model of the world in its hierarchical memory. Thoughts are what occur when this model runs on its own; memory recall leads to predictions, which act like sensory inputs, which lead to new memory recall, and so on. Our most contemplative thoughts are not driven by or even connected to the real world; they are purely a creation of our model. We close our eyes and seek quiet so that our thinking will not be interrupted by sensory input. Of course our model was originally created by exposure to the real world through our senses, but when we plan and think about the world, we do so via the cortical model, not the world itself.

To the cortex, our bodies are just part of the external world. Remember, the brain is in a quiet and dark box. It knows about the world only via the patterns on the sensory nerve fibers. From the brain's perspective as a pattern device, it doesn't know about your body any differently than it knows about the rest of the world. There isn't a special distinction between where the body ends and the rest of the world begins. But the cortex has no ability to model the brain itself because there are no senses in the brain. Thus we can see why our thoughts appear independent of our bodies, why it feels like we have an independent mind or soul. The cortex builds a model of your body but it can't build a model of the brain itself. Your thoughts, which are located in the brain, are physically separate from the body and the rest of the world. Mind is independent of body, but not of brain.

We can clearly see this differentiation through trauma and disease. If someone loses a limb, his brain's model of the limb may nevertheless remain intact, resulting in a so-called phantom limb, which he can still feel attached to his body. On the flip side, if he suffers cortical trauma he may lose his model of the arm even though he retains the arm itself. In this case he may suffer what's known as alien limb syndrome and have the uncomfortable, perhaps intolerable, feeling that the arm is not his own and is being controlled by someone else. Some even insist that the limb should be amputated! If our brain stays intact while the rest of our body becomes ill, we have the feeling of a healthy mind trapped in a dying body, although what we really have is a healthy brain trapped in a dying body. It is natural to imagine that our mind will continue after the death of our body, but when the brain dies so does the mind. The truth of this is evident if our brains fail before our bodies. People with Alzheimer's disease or with serious brain damage lose their minds even if their bodies stay healthy.

WHAT IS IMAGINATION?

Conceptually, imagination is rather simple. Patterns flow into each cortical area either from your senses or from lower areas of the memory hierarchy. Each cortical area creates predictions, which are sent back down the hierarchy. To imagine something, you merely let your predictions turn around and become inputs. Without physically doing anything, you can follow the consequences of your predictions. "If this happens, then this will happen, then this will happen," and so on. We do this when preparing for a business meeting, playing a game of chess, preparing for a sports event, or doing a thousand other things.

In chess you imagine moving your knight to a certain position and then visualize what the board will look like after the move. With this image in mind, you predict what your opponent will do and what the board will look like following that move. Then you predict what you will do, and so on. You walk through the imagined steps and their consequences. Ultimately you decide, based on this imagined sequence of events, whether the initial move was a good one or not. Certain athletes, such as downhill skiers, can improve their performance if they mentally rehearse the racecourse over and over in their head. By closing their eyes and imagining each and every turn, every obstacle, and even being on the winning stand, they increase their chances of success. Imagining is just another word for planning. This is where the predictive ability of our cortex pays off. It permits us to know what the consequences of our actions will be before we do them.

Imagining requires a neural mechanism for turning a prediction into an input. In chapter 6 I proposed that cells in layer 6 are where precise prediction occurs. Cells in this layer project down to lower levels of the hierarchy, but they also project back up to the input cells in layer 4. Thus a region's outputs can become its own inputs. As I mentioned earlier, longtime cortical modeler Stephen Grossberg calls this circuit for imagination "folded feedback." If you close your eyes and imagine a hippopotamus, the visual area of your cortex will become active, just as it would if you actually were looking at a hippo. You see what you imagine.

WHAT IS REALITY?

People ask, with an expression of worry and astonishment, "Do you mean to say that our brains create a model of the world? And that the model can be more important than actual reality?"

"Well, yes, to some extent that would be true," I say.

"But doesn't the world exist outside my head?"

Of course it does. People are real, trees are real, my cat is real, the social situations you find yourself in are real. But your understanding of the world and your responses to it are based on predictions coming from your internal model. At any moment in time, you can directly sense only a tiny part of your world. That tiny part dictates what memories will be invoked, but it isn't sufficient on its own to build the whole of your current perception. For example, I am typing in my office right now and hear a knock on my front door. I know my mother has come to visit and I imagine she is downstairs, even though I haven't actually seen or heard her. There was nothing in the sensory input that was specifically tied to my mother. It is my memory model of the world that predicts she is here by analogy to past experience. Most of what you perceive is not coming through your senses; it is generated by your internal memory model.

So the question "What is reality?" is largely a matter of how accurately our cortical model reflects the true nature of the world.

Many aspects of the world around us are so consistent that nearly every human has the same internal model of them. As a baby, you learned that the light falling on a round object produces a certain shadow, and that you can assess the shape of most objects by cues from the natural world. You learned that if you flung a cup off your highchair, gravity always pulled it to the floor. You learned textures, geometry, colors, and the rhythms of day and night. The simple physical properties of the world are learned consistently by all people.

But much of our world model is based on custom, culture,

and what our parents teach us. These parts of our model are less consistent and might be totally different for different people. A child who is raised in a loving, caring home with parents who respond to his or her emotional needs will probably grow to adulthood predicting that the world is safe and loving. Children abused by one or both parents are more likely to see future events as dangerous and cruel, and believe that no one is to be trusted—no matter how well they are treated later. Much of psychology is based on the consequences of early life experience, attachment, and nurturance because that is when the brain first lays down its model of the world.

Your culture thoroughly shapes your world model. For example, studies show that Asians and Westerners perceive space and objects differently. Asians attend more to the space between objects, whereas Westerners mostly attend to objects—a difference that translates into separate aesthetics and ways of solving problems. Research has shown that some cultures, such as tribes in Afghanistan and some communities in the American South, are built on principles of honor and, as a result, are more likely to accept the naturalness of violence. Differing religious beliefs learned early in life can lead to completely different models of morality, how men and women are to be treated, and even the value of life itself. Clearly these differing models of the world can't all be correct in some absolute, universal way, even though they may seem correct to an individual. Moral reasoning, both the good and the bad, is learned.

Your culture (and family experience) teaches you stereotypes, which are unfortunately an unavoidable part of life. Throughout this book, you could substitute the word *stereotype* for *invariant memory* (or *invariant representation*) without substantially altering the meaning. Prediction by analogy is pretty much the same as judgment by stereotype. Negative stereotyping has terrible social consequences. If my theory of intelligence is right, we cannot rid people of their propensity to think in stereotypes, because stereotypes are how the cortex works. Stereotyping is an inherent feature of the brain.

The way to eliminate the harm caused by stereotypes is to teach our children to recognize false stereotypes, to be empathetic, and to be skeptical. We need to promote these critical-thinking skills in addition to instilling the best values we know. Skepticism, the heart of the scientific method, is the only way we know how to ferret out fact from fiction.

By now, I hope I have convinced you that mind is just a label of what the brain does. It isn't a separate thing that manipulates or coexists with the cells in the brain. Neurons are just cells. There is no mystical force that makes individual nerve cells or collections of nerve cells behave in ways that differ from what they would normally do. Given this fact, we can now turn our attention to how we might implement the ability of brain cells to remember and predict—our cortical algorithm—in silicon.

8

THE FUTURE OF INTELLIGENCE

It's hard to predict the ultimate uses of a new technology. As we've seen throughout this book, brains make predictions by analogy to the past. So our natural inclination is to imagine that a new technology will be used to do the same kinds of things as a previous technology. We imagine using a new tool to do something familiar, only faster, more efficiently, or more cheaply.

Examples are abundant. People called the railroad the "iron horse" and the automobile the "horseless carriage." For decades the telephone was viewed in the context of the telegraph, something that should be used only to communicate important news or emergencies; it wasn't until the 1920s that people started using it casually. Photography was at first used as a new form of portraiture. And motion pictures were conceptualized as a variation on stage plays, which is why movie theaters had retracting curtains over the screens for much of the twentieth century.

Yet the ultimate uses of a new technology are often unexpected and more far-reaching than our imaginations can at first grasp. The telephone has evolved into a wireless voice and data communications network permitting any two people on the planet to communicate with each other, no matter where they are, via voice, text, and images. The transistor was invented by Bell Labs in 1947. It was instantly clear to people that the device was a breakthrough, but the initial applications were just improvements on old applications: transistors replaced vacuum tubes. This led to smaller and more reliable radios and computers, which was important and exciting in its day, but the main differences were the size and reliability of the machines. The transistor's most revolutionary applications weren't discovered until later. A period of gradual innovation was necessary before anyone could conceive of the integrated circuit, the microprocessor, the digital signal processor, or the memory chip. The microprocessor, likewise, was first developed, in 1970, with desktop calculators in mind. Again, the first applications were just replacements of existing technologies. The electronic calculator was a replacement for the mechanical desktop calculator. Microprocessors were also clear candidates to replace the solenoids that were then used in certain kinds of industrial control, such as switching traffic lights. However, it was years before the true power of the microprocessor began to be manifest. No one at the time could foresee the modern personal computer, the cell phone, the Internet, the Global Positioning System, or any other piece of today's bread-and-butter information technology.

By the same token, we would be foolish to think we can predict the revolutionary applications of brainlike memory systems. I fully expect these intelligent machines to improve life in all sorts of ways. We can be sure of it. But predicting the future of technology more than a few years out is impossible. To appreciate this you need only read some of the absurd prognostications futurists have confidently made over the years. In the 1950s, it was predicted that by the year 2000 we'd all have atomic reactors in our basements and take our vacations on the moon. But as long as we keep these cautionary tales in mind, there's a lot to be gained by speculating about what intelligent machines will be like. At a minimum, there are certain broad and useful conclusions we can draw about the future.

The questions are intriguing ones. Can we build intelligent machines, and, if so, what will they look like? Will they be closer to the humanlike robots seen in popular fiction, the black or beige box of a personal computer, or something else? How will they be used? Is this a dangerous technology that can harm us or threaten our personal liberties? What are the obvious applications for intelligent machines, and is there any way we can know what the fantastic applications will be? What will the ultimate impact of intelligent machines be on our lives?

CAN WE BUILD INTELLIGENT MACHINES?

Yes, we can build intelligent machines, but they may not be what you expect. Although it may seem like the obvious thing to do, I don't believe we will build intelligent machines that act like humans, or even interact with us in humanlike ways.

One popular notion of intelligent machines comes to us from movies and books—they are the lovable, evil, or occasionally bumbling humanoid robots that converse with us about feelings, ideas, and events, and play a role in endless science-fiction plots. A century of science fiction has trained people to view robots and androids as an inevitable and desirable part of our future. Generations have grown up with images of Robbie the Robot from *Forbidden Planet,* R2D2 and C3PO from *Star Wars,* and Lieutenant Commander Data from *Star Trek*. Even HAL in the movie *2001: A Space Odyssey,* although not possessing a body, was very humanlike, designed to be as much a companion as a programmed copilot for the humans on their long space journey. Limited-application robots—things like smart cars, autonomous minisubmarines to explore the deep ocean, and self-guided vacuum cleaners or lawn mowers—are feasible and may well grow more common someday. But androids and robots like Commander Data and C3PO are going to remain fictional for a very long time. There are a couple of reasons for this.

First, the human mind is created not only by the neocortex but also by the emotional systems of the old brain and by the complexity of the human body. To be human you need all of your biological machinery, not just a cortex. To converse like a human on all matters (to pass the Turing Test) would require an intelligent machine to have most of the experiences and emotions of a real human, and to live a humanlike life. Intelligent machines will have the equivalent of a cortex and a set of senses, but the rest is optional. It might be entertaining to watch an intelligent machine shuffle around in a humanlike body, but it will not have a mind that is remotely humanlike unless we imbue it with humanlike emotional systems and humanlike experiences. That would be extremely difficult and, it seems to me, quite pointless.

Second, given the cost and effort that would be necessary to build and maintain humanoid robots, it is difficult to see how they could be practical. A robot butler would be more expensive and less helpful than a human assistant. While the robot might be "intelligent," it would not have the kind of rapport and easy understanding a human assistant would have by virtue of being a fellow human being.

Both the steam engine and the digital computer evoked robotic visions, which never came to fruition. Similarly, when we think of building intelligent machines, many people find it natural to imagine humanlike robots once again, but it is unlikely to happen. Robots are a concept born of the industrial revolution and refined by fiction. We should not look to them for inspiration in developing genuinely intelligent machines.

So what will intelligent machines look like if not walking talking robots? Evolution discovered that if it attached a hierarchical memory system to our senses, the memory would model the world and predict the future. Borrowing from nature, we should build intelligent machines along the same lines. Here, then, is the recipe for building intelligent machines. Start with a set of senses to extract patterns from the world. Our intelligent machine may have a set of senses that differ from a human's, and may even "exist" in a world unlike our own (more on this later). So don't assume that it has to have a set of eyeballs and a pair of ears. Next, attach to these senses a hierarchical memory system that works on the same principles as the cortex. We will then have to train the memory system much as we teach children. Over repetitive training sessions, our intelligent machine will build a model of *its* world as seen through *its* senses. There will be no need or opportunity for anyone to program in the rules of the world, databases, facts, or any of the high-level concepts that are the bane of artificial intelligence. The intelligent machine must learn via observation of its world, including input from an instructor when necessary. Once our intelligent machine has created a model of its world, it can then see analogies to past experiences, make predictions of future events, propose solutions to new problems, and make this knowledge available to us.

Physically, our intelligent machine might be built into planes or cars, or sit stoically on a rack in a computer room. Unlike humans, whose brains must accompany their bodies, the memory system of an intelligent machine might be located remotely from its sensors (and "body," if it had one). For example, an intelligent security system might have sensors located throughout a factory or a town, but the hierarchical memory system attached to those sensors could be locked in a basement of one building. Therefore, the physical embodiment of an intelligent machine could take many forms.

There is no reason that an intelligent machine should look, act, or sense like a human. What makes it intelligent is that it can understand and interact with its world via a hierarchical memory model and can think about its world in a way analogous to how you and I think about our world. As we will see, its thoughts and actions might be completely different from anything a human does, yet it will still be intelligent. Intelligence is measured by the predictive ability of a hierarchical memory, not by humanlike behavior.

Let's turn our attention to the largest technical challenge we will face when building intelligent machines, creating the memory. To build intelligent machines, we will need to construct large memory systems that are hierarchically organized and that work like the cortex. We will confront challenges with capacity and connectivity.

Capacity is the first issue. Let's say the cortex has 32 trillion synapses. If we represented each synapse using only two bits (giving us four possible values per synapse) and each byte has eight bits (so one byte could represent four synapses), then we would need roughly 8 trillion bytes of memory. A hard drive on a personal computer today has 100 billion bytes, so we would need about eighty of today's hard drives to have the same amount of memory as a human cortex. (Don't worry about the exact numbers because they are all rough guesses.) The point is, this amount of memory is definitely buildable in the lab. We aren't off by a factor of a thousand, but it is also not the kind of machine you could put in your pocket or build into your toaster. What is important is that the amount of memory required is not out of the question, whereas only ten years ago it would have been. Helping us is the fact that we don't have to recreate an entire human cortex. Much less may suffice for many applications.

Our intelligent machines will need lots of memory. We will probably start building them using hard drives or optical disks, but eventually we will want to build them out of silicon as well.

Silicon chips are small, low power, and rugged. And it is only a matter of a time before silicon memory chips could be made with enough capacity to build intelligent machines. In fact, there is an advantage intelligent memory has over conventional computer memory. The economics of the semiconductor industry is based on the percentage of chips that have errors. For many chips even a single error will make the chip useless. The percentage of good chips is called the yield. It determines whether a particular chip design can be manufactured and sold at a profit. Because the chance of an error increases as the size of the chip does, most chips today are no bigger than a small postage stamp. The industry has boosted the amount of memory on a single chip not by making the chip larger but, mostly, by making the individual features on the chip smaller.

But intelligent memory chips will be inherently tolerant of faults. Remember, no single component of your brain holds any indispensable item of data. Your brain loses thousands of neurons each day, yet your mental capacity decays at only a slow pace throughout your adult life. Intelligent memory chips will work on the same principles as cortex, so even if a percentage of the memory elements come out defective, the chip will still be useful and commercially viable. Most likely, the inherent tolerance to errors of brainlike memory will allow designers to build chips that are significantly larger and denser than today's computer memory chips. The result is that we may be able to put a brain in silicon sooner than current trends might indicate.

The second problem we have to overcome is *connectivity*. Real brains have large amounts of subcortical white matter. As we noted earlier, the white matter is made up of the millions of axons streaming this way and that just beneath the thin cortical sheet, connecting the different regions of the cortical hierarchy with each other. An individual cell in the cortex may connect to five or ten thousand other cells. This kind of massively parallel wiring is difficult or impossible to implement using traditional silicon manufacturing techniques. Silicon chips are made by depositing a few layers of metal, each separated by a layer of insulation. (This process has nothing to do with the layers in the cortex.) The layers of metal contain the "wires" of the chip, and because wires can't cross within a layer, the total number of wired connections is limited. This is not going to work for brainlike memory systems, where millions of connections are necessary. Silicon chips and white matter are not very compatible.

A lot of engineering and experimentation will be necessary to solve this problem, but we know the basics of how it will be solved. Electrical wires send signals much more quickly than the axons of neurons. A single wire on a chip can be shared, and therefore used for many different connections, whereas in the brain each axon belongs to just one neuron.

A real-world example is the telephone system. If we ran a line from every telephone to every other telephone, the surface of the globe would be buried under a jungle of copper wire. What we do instead is have all the telephones share a relatively small number of high-capacity lines. This method works as long as the capacity of each line is far greater than the capacity required to transmit a single conversation. The telephone system meets this requirement: a single fiber optic cable can carry a million conversations at once.

Real brains have dedicated axons between all cells that talk to each other, but we can build intelligent machines to be more like the telephone system, sharing connections. Believe it or not, some scientists have been thinking about how to solve the brain chip connectivity problem for many years. Even though the operation of the cortex remained a mystery, researchers knew that we would someday unravel the puzzle, and then we would have to face the issue of connectivity. We don't need to review the different approaches here. Suffice it to say that connectivity might be the biggest technical obstacle we face in building intelligent machines but we should be able to handle it.

Once the technological challenges are met, there are no fundamental problems that prevent us from building genuinely intelligent systems. Yes, there are lots of issues that will need to be addressed to make these systems small, low cost, and low power, but nothing is standing in our way. It took fifty years to go from room-size computers to ones that fit in your pocket. But because we are starting from an advanced technological position, the same transition for intelligent machines should go much faster.

SHOULD WE BUILD INTELLIGENT MACHINES?

Throughout the twenty-first century, intelligent machines will emerge from the realm of science fiction into fact. Before we get there, we should think about ethical issues, whether the possible dangers of intelligent machines outweigh the likely benefits.

The prospect of machines that can think and act on their own has worried people for a long time. This is understandable. New areas of knowledge and new technologies always scare people when they first come along. Human creativity lets us imagine the terrible ways a new technology may take over our bodies, outmode our usefulness, or cancel out the very value of human life. But history shows that these dark imaginings almost never play out the way we expect. When the industrial revolution came along, we feared electricity (remember Frankenstein?) and steam engines. Machinery that had its own energy, that could move itself in complex ways, seemed miraculous and at the same time potentially sinister. But electricity and internal combustion engines are no longer strange and sinister. They are as much a part of our environment as air and water.

When the information revolution began, we quickly came to fear computers. There were countless science-fiction stories about powerful computers or computer networks that spontaneously became self-aware and then turned on their organic masters. But now that computers have become integrated into daily life, this fear seems absurd. The computer in your home, or the Internet, has as much chance of spontaneously turning sentient as does a cash register.

Any technology can be applied to good or evil ends, of course, but some are more inherently prone to misuse or catastrophe than others. Atomic energy is dangerous whether it's in the form of nuclear warheads or power plants because a single accident or a single misuse could harm or kill millions of people. And although nuclear energy is valuable, alternatives are available. Vehicular technology can take the form of tanks and fighter jets, or it can take the form of cars and passenger airplanes, and a mishap or misuse can cause harm to many people. But vehicles are arguably both more essential to modern life and less dangerous than nuclear power. The damage caused by a single misuse of an airplane is much less than that of a nuclear bomb. There are many technologies that are almost wholly beneficial. Telephones are an example. Overwhelmingly, their tendency to bring and keep people together exceeds any negative effects. The same goes for electricity and public health science. In my opinion, intelligent machines are going to be one of the least dangerous, most beneficial technologies we have ever developed.

Still, some thinkers, like Sun Microsystems cofounder Bill Joy, fear that we may develop intelligent robots that could escape our control, swarm the Earth, and remake it according to their own agenda. The image puts me in mind of those magically animated broomsticks from the *Sorcerer's Apprentice,* regenerating themselves from their splinters and working tirelessly to bring about disaster. Along similar lines, some AI optimists offer extended-life prophecies that are unsettling. For instance, Ray Kurzweil talks about the day when nanorobots will crawl within your brain, recording every synapse and every connection, and then report all the information to a supercomputer, which will reconfigure itself into you! You'll become a "software" version of yourself that will be practically immortal. These two predictions about machine intelligence, the intelligent machines–run-amok scenario and the upload-your-brain-into-a-computer scenario, seem to surface over and over.

Building intelligent machines is not the same as building self-replicating machines. There is no logical connection whatsoever between them. Neither brains nor computers can directly self-replicate, and brainlike memory systems will be no different. While one of the strengths of intelligent machines will be our ability to mass-produce them, that's a world apart from self-replication in the manner of bacteria and viruses. Self-replication does not require intelligence, and intelligence does not require self-replication.

Further, I seriously doubt we will ever be able to copy our minds into machines. There are at present, as far as I know, no actual or imagined methods capable of recording the trillions of details that make "you." We would need to record and re-create all of your nervous system and your body, not just your neocortex. And we would need to understand how all of it works. One day, certainly, we might be to able do this, but the challenges extend far beyond understanding how the cortex works. Figuring out the neocortical algorithm and building it into machines from scratch is one thing, but scanning in the zillions of operational details of a living brain and replicating them in a machine is something completely different.

Beyond self-replication and the copying of minds, people have another concern with intelligent machines. Might intelligent machines somehow threaten large portions of the population, as nuclear bombs do? Might their presence lead to the superempowerment of small groups or malevolent individuals? Or might the machines become evil and work against us, like the implacable villains in *The Terminator* or the *Matrix* movies?

The answer to these questions is no. As information devices, brainlike memory systems are going to be among the most useful technologies we have yet developed. But like cars and computers, they will only be tools. Just because they are going to be intelligent does not mean they will have special abilities to destroy property or manipulate people. And just as we wouldn't put the control of the world's nuclear arsenal under the authority of one person or one computer, we will have to be careful not to rely too much on intelligent machines, for they will fail as all technology does.

This gets us to the malevolence question. Some people assume that being intelligent is basically the same as having human mentality. They fear that intelligent machines will resent being "enslaved" because humans hate being enslaved. They fear that intelligent machines will try to take over the world because intelligent people throughout history have tried to take over the world. But these fears rest on a false analogy. They are based on a conflation of intelligence—the neocortical algorithm—with the emotional drives of the old brain—things like fear, paranoia, and desire. But intelligent machines will not have these faculties. They will not have personal ambition. They will not desire wealth, social recognition, or sensual gratification. They will not have appetites, addictions, or mood disorders. Intelligent machines will not have anything resembling human emotion unless we painstakingly design them to. The strongest applications of intelligent machines will be where the human intellect has difficulty, areas in which our senses are inadequate, or in activities we find boring. In general, these activities have little emotional content.

Intelligent machines will range from simple, single-application systems to very powerful superhuman intelligent systems, but unless we go out of our way to make them humanlike, they won't be. Maybe someday we will have to place restrictions on what people can do with intelligent machines, but that day is a long way off, and when it comes, the ethical issues are likely to be relatively easy compared with such present-day moral questions as those surrounding genetics and nuclear technology.

WHY BUILD INTELLIGENT MACHINES?

Now on to the question What will intelligent machines do?

I'm often asked to give talks about the future of mobile computing. A conference organizer will ask me to describe what handheld computers or cell phones will look like in five or in twenty years. They want to hear my vision for the future. I can't do this. I try to avoid what I call the V word altogether. To make a point, I once walked onstage with a wizard's hat and a crystal ball. I explained that no one can see the future in detail. Crystal balls are fiction, and anyone who pretends to know exactly what will happen in the coming years is sure to fail. Instead, the best we can do is to understand broad trends. If you understand a broad idea, you can successfully pursue it wherever it goes as the details unfold.

The most famous example of a technology trend is Moore's Law. Gordon Moore correctly predicted that the number of circuit elements that could be placed on a silicon chip would double every year and a half. Moore didn't say whether the chips would be memory chips, central processing units, or something else. He didn't say what kinds of products the chips would be used in. He didn't predict whether the chips would be housed in plastic or ceramic or glued to circuit boards. He didn't say anything about the various processes used to manufacture chips. He stuck to the broadest trend he could, and he got it right.

We in the present day cannot predict the ultimate uses of intelligent machines. There is simply no way we are going to get the details right. If I, or anyone else, predict in detail what these machines will do, we will inevitably be proven wrong. Nonetheless, we can do better than just shrugging our shoulders. There are two lines of thought that may be helpful. One is to envision the very near-term uses for brainlike memory systems—the obvious, but less interesting, things to try first. The second approach is to think about the long-term trends, like Moore's Law, that can help us imagine the applications that could possibly be part of our future.

Let's begin with some near-term applications. These are the things that seem obvious, like replacing tubes in a radio with transistors or building calculators with a microprocessor. And we can start by looking at some areas that AI tried to tackle but couldn't solve—speech recognition, vision, and smart cars.

If you have ever tried to use speech recognition software to enter text on a personal computer, you know how dumb it can be. Like Searle's Chinese Room, the computer has no understanding of what is being said. The few times I tried these products, I grew frustrated. If there was any noise in the room, from a dropped pencil to someone speaking to me, extra words would appear on my screen. The recognition error rates were high. Often the words the software thought I said made no sense. "Remember to fell Mary that the bog is ready to be piqued up." A child would know this is wrong, but not the computer. Similarly, so-called natural language interfaces have been a goal for computer scientists for years. The idea is for you to be able to tell a computer or other appliance what it is you want, in plain language, and let the machine do the work. On a personal digital assistant, or PDA, you might say, "Move my daughter's basketball game on Sunday to ten in the morning." This kind of thing has been impossible to do well with traditional AI. Even if the computer could recognize each word, for it to complete the task it might need to know where your daughter goes to school, that you probably mean the coming Sunday, and maybe what a basketball game is because the appointment might only say "Menlo vs. St. Joe." Or perhaps you want a computer to listen to the audio stream from a radio broadcast for any mention of a particular product, but the radio announcer describes the product without using its name. You and I would know what she is talking about, but not a computer.

These and many other applications require that a machine be able to listen to spoken language. But computers can't perform these tasks, because they do not understand what is being said. They match auditory patterns to word templates by rote, without knowing what the words mean. Imagine if you learned to recognize the sounds of individual words in a foreign language, but not the meaning of the words, and I asked you to transcribe a conversation in that language. As the conversation flows, you have no idea what it is about, but you try to pick out the words in isolation. However, the words overlap and interfere, and pieces of sound drop out because of noise. You would find it extremely difficult to separate words and recognize them. These obstacles are what speech recognition software struggles with today. Engineers have discovered that by using probabilities of word transitions, they can improve the software's accuracy somewhat. For example, they use rules of grammar to decide between homonyms. This is a very simple form of prediction, but the systems are still dumb. Today's speech recognition software succeeds only in highly constrained situations in which the number of words you might say at any given moment is limited. Yet humans perform many language-related tasks easily, because our cortex understands not only words but sentences and the context within which they are spoken. We anticipate ideas, phrases, and individual words. Our cortical model of the world does this automatically.

So we can expect that cortexlike memory systems will transform fallible speech recognition into robust speech understanding. Instead of programming in probabilities for single word transitions, a hierarchical memory will track accents, words, phrases, and ideas and use them to interpret what is being said. Like a person, such an intelligent machine could distinguish between various speech events—for example, a discussion between you and a friend in the room, a phone conversation, and editing commands for a book. It won't be easy to build these machines. To fully understand human language, a machine will have to experience and learn what humans do. So even though it may be many years before we can build an intelligent machine that understands language as well as you and I do, we will be able, in the near term, to improve on the performance of existing speech recognition systems by using cortical-like memories.

Vision offers another set of applications that AI has been unable to accomplish but that real intelligent systems should be able to handle. Today there is no machine that can look at a natural scene—the world in front of your eyes, or a picture using a camera—and describe what it sees. There are a few successful applications of machine vision that work in very restricted domains, such as visually aligning chips on a circuit board or matching facial features to a database, but it's currently impossible for a computer to identify varieties of objects or analyze a scene more generally. You have no difficulty looking around a room to find a place to sit, but don't ask a computer to do it. Imagine looking at a video screen from a security camera. Can you tell the difference between someone knocking on a door holding a gift and someone knocking down the door with a crow bar? Of course you can, but the distinction is well beyond the capability of today's software. Consequently we hire people to keep an eye on the screens of security cameras round-the-clock, looking for something suspicious. It is difficult for human watchers to stay alert, whereas an intelligent machine could perform the task tirelessly.

Finally, let's look at transportation. Cars are getting pretty sophisticated. They have Global Positioning Systems to map your route from A to B, sensors to turn on the lights when it gets dark, accelerometers to deploy airbags, and proximity sensors to tell you if you are about to back over something. There are even cars that can drive autonomously on special highways or in ideal conditions, although they are not commercially available. But to drive a car safely and effectively on all types of roads and in all kinds of traffic conditions requires more than a few sensors and feedback control circuits. To be a good driver, you must understand traffic, other drivers, the way cars work, signal lights, and a slew of other things. You need to be able to understand hazard signs or notice when another car is driving dangerously. You need to see a directional signal on another car and anticipate that it is likely to change lanes, or, if the signal has been on for several minutes, realize that the driver probably doesn't know that the signal is on and therefore probably won't change lanes. You need to recognize that a puff of smoke far ahead might mean that an accident has occurred and therefore you should slow down. A driver seeing a ball roll across the street and thinking, automatically, that a child may run out to grab it intuitively comes to a stop.

Let's say we wanted to build a truly smart car. The first thing we would do is to select a set of sensors that would allow our smart car to sense its world. We might start with a camera for seeing, perhaps multiple cameras in front and back, and microphones for hearing, but we might also want to provide radar or ultrasound sensors that can accurately determine the range and speed of other objects in lit or dark conditions. The point is, we don't have to rely on or restrict ourselves to the senses humans use. The cortical algorithm is flexible, and, as long as we design our hierarchical memory system appropriately, it should work no matter what types of sensors we install. Our car could, in theory, be better at sensing the world of traffic than we are because its set of senses can be chosen to match the task. The sensors would then be attached to a sufficiently large hierarchical memory system. The car designers would train the smart car's memory by exposing it to real-world conditions so that it learns to build a model of its world in the same way humans do—only in a more limited realm. (For example, the car needs to know about roads but not about elevators and airplanes.) The car's memory would learn the hierarchical structure of traffic and roads so that it could understand and anticipate what is likely to happen in its world of moving automobiles, road signs, obstacles, and intersections. The car's engineers could design the memory system so that it actually drives the car or just monitors what happens when you drive. It could give advice, or take over in extreme situations, like a backseat driver that you don't resent. Once the memory is fully trained and the car can understand and handle anything that might happen, the engineers would have the option of permanently setting the memory so that all cars coming off the assembly line behave the same way, or they could design the memory to continue learning after the car is sold. And as with a computer but not a human, the memory could be reprogrammed with an updated version should new conditions warrant doing so.

I am not saying we will definitely build smart cars or machines that understand language and vision. But these are good examples of the kinds of devices we might research and develop, and that seem possible to build.

Personally, I am less interested in the obvious applications of intelligent machines. To me, the true benefit and excitement of a new technology is in finding uses for it that were inconceivable before. In what ways will intelligent machines surprise us, and what fantastic capabilities will emerge over time? I am certain that hierarchical memories, like the transistor and the microprocessor, will transform our lives for the better in incredible ways, but how? One way we can glimpse the future of intelligent machines is to think of aspects of the technology that will scale well. That is, which attributes of intelligent machines will grow cheaper and cheaper, faster and faster, or smaller and smaller. Things that grow at exponential rates rapidly outpace our imagination and are most likely to play a key role in the most radical evolutions in future technology.

Examples of technologies that have improved exponentially for many years include the silicon memory chip, the hard disk, DNA sequencing techniques, and fiber optics. These rapidly scaling technologies have been the basis for many new products and businesses. In a different fashion, software also scales well. A desirable program, once it is written, can be copied endlessly at virtually no cost.

In contrast, some technologies, such as batteries, motors, and traditional robotics, scale poorly. Despite lots of effort and steady improvements, a robotic arm built today is not dramatically better than one built a few years ago. The developments in robotics are gradual and modest, nothing like the exponential growth curves of chip design or software proliferation. A robot arm built in 1985 for a million dollars cannot be built today a thousand times as strong for just ten dollars. Likewise, batteries today are not vastly better than they were ten years ago. You might say they are two or three times better, but not a thousand or ten thousand times better, and progress moves forward only bit by bit. If the capacity of batteries increased at the same rate as the capacity of hard disks, then cell phones and other electronics would never have to be charged, and lightweight electric cars that traveled a thousand miles per charge would be common.

So it behooves us to think about which aspects of brainlike memory systems will scale dramatically beyond our biological brains. These attributes will suggest where the technology will ultimately land. I see four attributes that will exceed our own abilities: speed, capacity, replicability, and sensory systems.

Speed

While neurons work on the order of milliseconds, silicon operates on the order of nanoseconds (and is still getting faster). That's a million-fold difference, or six orders of magnitude.

The speed difference between organic and silicon-based minds will be of great consequence. Intelligent machines will be able to think as much as a million times faster than the human brain. Such a mind could read whole libraries of books or study huge, complicated bodies of data—tasks that would take you or me years to complete—in mere minutes, while getting exactly the same understanding out of it. There is nothing magic about this. Biological brains evolved with two time-related constraints. One is the speed at which cells can do things and the other is the speed at which the world changes. It might not be too useful for a biological brain to think a million times faster if the world around it is inherently slow. But there is nothing about the cortical algorithm that says it must always operate slowly. If an intelligent machine conversed or interacted with a human, it would have to slow down to work at human speed. If it read a book by flipping pages, there would be a limit to how fast it could read. But when it is interfacing with the electronic world, it could function much more quickly. Two intelligent machines could hold a conversation a million times faster than two humans. Imagine the progress of an intelligent machine that solved mathematical or scientific problems a million times faster than a human. In ten seconds it could give as much thought to a problem as you could in a month. Never-tiring, never-bored minds of such lightning speed are sure to be useful in ways we can't yet imagine.

Capacity

Despite the impressive memory capacity of a human cortex, intelligent machines can be built to surpass it by large margins. The size of our brains has been constrained by several biological factors, including the ratio of infant skull size to maternal pelvis diameter, the high metabolic cost of running a brain (your brain is about 2 percent of your body weight but uses around 20 percent of the oxygen you breathe), and the slow speed of neurons. But we can build intelligent memory systems of any size, and, unlike the blind, meandering process of evolution, we can bring foresight and specific intention to the details of the design. The capacity of the human neocortex may turn out to be relatively modest some decades from now.

As we build intelligent machines, we could increase their memory capacity in several ways. Adding depth to the hierarchy will lead to deeper understanding—the ability to see higher-order patterns. Enlarging the capacity within regions will allow the machine to remember more details, or perceive with greater acuity, in the same way a blind person has a more refined sense of touch or hearing. And adding new senses and sensory hierarchies permits the device to construct better models of the world, as I will discuss shortly.

It will be interesting to see if there is an upper limit to how big an intelligent memory system can get and in what dimensions. Conceivably, a device might become too cluttered to be useful, or it might fail as it approaches some theoretical limit. Perhaps the human brain is already near the maximum theoretical size, but I find that unlikely. Human brains became large very recently in evolutionary time, and there is nothing to suggest that we are at some stable maximum size. Whatever the peak capacity for an intelligent memory system turns out to be, the human brain is almost certainly not there. It's probably not even close.

One way to see what these systems might do is to look at the limits of known human performance. Einstein was undoubtedly extremely smart, but his brain was still a brain. We can assume his extraordinary intelligence was largely the product of physical differences between his brain and the typical human brain. What made Einstein so rare was that the human genome doesn't often produce brains like his. However, when we are designing brains in silicon, we can build them any way we want. They could all be capable of Einstein's high level of thought, or even smarter. At another extreme, savants can give us insight into other possible dimensions of intelligence. Savants are mentally retarded individuals who exhibit remarkable abilities such as near-photographic memories or the capacity to perform difficult mathematical calculations at lightning speeds. Their brains, although not typical, are still brains, working with the cortical algorithm. If an atypical brain can have amazing memory abilities, then, in theory, we could add those capabilities to our artificial brains. Not only do these extremes of human mental ability indicate what should be possible to re-create; they represent the directions in which we can probably exceed the best human performance.

Replicability

Each new organic brain must be grown and trained de novo, a process that takes decades in human beings. Each human must discover for himself or herself the basics of coordinating the body's limbs and muscle groups, of balancing and moving, and learn the general properties of multitudinous objects, animals, and other people; the names of things and the structure of the language; and the rules of family and society. Once those basics are mastered, years and years of formal schooling commence. Every individual must trudge up the same set of learning curves in life, even though they have been trudged countless times by others, all to build a model of the world in the cortex.

Intelligent machines need not undergo this long learning curve, since chips and other storage can be replicated endlessly and the contents transferred easily. In this sense, intelligent machines could replicate like software. Once a prototype system has been satisfactorily honed and trained, it could be copied as many times as we please. It may take years of chip design, hardware configuration, training, and trial and error to perfect the memory system for a smart car, but once the final product is achieved, it can be mass-produced. As I mentioned earlier, we could choose to allow the copies to continue learning or not. For some applications, we will want to limit our intelligent machines to operate in a tested and known way. Once a smart car knows everything we need it to know, we wouldn't want it to develop bad habits or come to believe in some false analogy it thinks it sees. If nothing else, we would expect all cars of a similar make to behave similarly. But for other applications, we'll want our brainlike memory systems to be fully capable of continuous learning. For instance, an intelligent machine designed to discover mathematical proofs will need the ability to learn from experience, to apply old insights to new problems, and to be generally flexible and open-ended.

It should be possible to share components of learning the way we share components of software. An intelligent machine of a particular design could be reprogrammed with a new set of connections to lead to different behavior, as if I could download a new set of connections into your brain and instantly change you from an English speaker to a French speaker, or from a political science professor to a musicologist. People could swap and build on the work of others. Let's say I have developed and trained a machine with a superior visual system, and another person has developed and trained a machine with a superior sense of hearing. With the proper design, we could combine the best of both systems without retraining from the bottom up. Sharing expertise in this way is simply impossible for humans. The business of building intelligent machines could evolve along the same lines as the computer industry, with communities of people training intelligent machines to have specialized knowledge and abilities, and selling and swapping the resultant memory configurations. Reprogramming an intelligent machine will not be too different from running a new video game or installing a piece of software.

Sensory Systems

Humans have a handful of senses. These senses are deeply ingrained in our genes, in our bodies, and in the subcortical wiring of our brains. We can't change them. Sometimes we use technology to augment our senses, such as with night-vision goggles, radar, or the Hubble space telescope. These high-tech instruments are clever tricks of data translation, not new modes of perception. They convert information we can't sense into visual or auditory displays that we can interpret. Still, it is a tribute to our brain's amazing flexibility that we can look at a radar screen and understand what it represents. Many animal species possess truly different senses, such as the echolocation of bats and dolphins, the ability of bees to see polarized and ultraviolet light, and the electric field sense of some fish.

Our intelligent machines could perceive the world through any sense found in nature as well as new senses of purely human design. Sonar, radar, and infrared vision are obvious examples of the kinds of nonhuman senses that we may want our intelligent machines to possess. But they are only the beginning.

Far more interesting is the way intelligent machines could experience genuinely exotic, alien worlds of sensation. As we have seen, the neocortical algorithm is fundamentally concerned with finding patterns in the world. It has no preference for the physical origins of those patterns. So long as the inputs to the cortex are nonrandom and have a certain amount of richness or statistical structure, an intelligent system will form invariant memories and predictions about them. There is no reason for these input patterns to be analogous to animal senses, or even to derive from the real world at all. It is in the realm of exotic senses that, I suspect, the revolutionary uses of intelligent machines lie.

For example, we could design a sensory system that spans the globe. Imagine weather sensors spaced every fifty or so miles across a continent. These sensors would be analogous to the cells in a retina. At any point in time, two adjacent weather sensors will have a high correlation in their activity, just like two adjacent cells on a retina. There are large weather objects, such as storms and fronts that move and change over time, just as there are visual objects that move and change over time. By attaching this sensory array to a large cortical-like memory, we would enable the system to learn to predict the weather in the way that you and I learn to recognize visualize objects and predict how they move over time. The system would see local weather patterns, large weather patterns, and patterns that exist over decades, years, or hours. By placing sensors close together in some regions, we could create the equivalent of a fovea, allowing our intelligent weather brain to understand and predict microclimates. Our weather brain would think about and understand global weather systems as you and I think about and understand objects and people. Meteorologists aim to do something similar today. They collect recordings from diverse locations and use supercomputers to simulate the climate and forecast the future. But this approach, which is fundamentally different from the way an intelligent machine would work, is akin to how a computer plays chess—dumb and without understanding—whereas our intelligent weather machine is akin to how a human plays chess—thoughtfully and with understanding. The intelligent weather machine would discover patterns that humans have not. It was only in the 1960s that the weather phenomenon known as El Niño was discovered. Our weather brain could find more patterns like El Niño, or learn how to predict tornadoes or monsoons far better than humans. Putting large amounts of weather data into a form that humans can readily understand is difficult; our weather brain, in contrast, would sense and think about weather directly.

Other large distributed sensory systems could allow us to build intelligent machines that understand and predict animal migrations, changes in demographics, and the spread of disease. Imagine having sensors distributed over a country's electrical power grid. An intelligent machine attached to these sensors would observe the ebb and flow of electricity consumption in the same way you and I see the ebb and flow of traffic on a road, or the movement of people at an airport. Through repeated exposure, humans learn to predict these patterns—just ask an employee who commutes by car, or an airport security guard. Similarly, our intelligent electrical grid monitor would be able to predict demands for power, or dangerous situations likely to lead to a power outage, better than a human. We might combine sensors for weather and for human demographics, in order to anticipate political unrest, famines, or disease outbreaks. Like a supersmart diplomat, intelligent machines may play a role in reducing conflict and human suffering. You might think intelligent machines would need emotions to foresee patterns involving human behavior, but I don't think so. We are not born with a set culture, a set of values, and a set religion; we learn them. And just as I can learn to understand the motivations of people with values different from mine, intelligent machines can comprehend human motivations and emotions, even if the machine doesn't have those emotions itself.

We could make senses that sample minute entities. It is theoretically possible to have sensors that could represent patterns in cells or large molecules. For example, an important challenge today is to understand how the shape of a protein molecule can be predicted from the sequence of amino acids that comprise the protein. Being able to predict how proteins fold and interact would accelerate the development of medicines and the cures for many diseases. Engineers and scientists have created three-dimensional visual models of proteins, in an effort to predict how these complex molecules behave. But try as we might, the task has proven too difficult. A superintelligent machine, on the other hand, with a set of senses specifically tuned to this question might be able to answer it. If this sounds far-fetched, remember that we wouldn't be surprised if humans could solve the problem. Our inability to tackle the issue may be related, primarily, to a mismatch between the human senses and the physical phenomena we want to understand. Intelligent machines can have custom senses and larger-than-human memory, enabling them to solve problems we can't.

With the proper senses and a slight restructuring of the cortical memory, our intelligent machines might live and think in virtual worlds used in mathematics and physics. For example, many endeavors in math and science require understanding how objects behave in worlds that have more than three dimensions. String theorists, who study the nature of space itself, think about the universe as having ten or more dimensions. Humans have great difficulty thinking about mathematical problems in four or more dimensions. Perhaps an intelligent machine of the correct design could understand high-dimensional spaces in the same way that you and I understand three-dimensional spaces, and therefore be adept at predicting how they behave.

Finally, we might unite a bunch of intelligent systems in a grand hierarchy, just as our cortex unites hearing, touch, and vision higher up the cortical hierarchy. Such a system would automatically learn to model and predict the patterns of thinking in populations of intelligent machines. With distributed communications mediums such as the Internet, the individual intelligent machines could be distributed around the globe. Larger hierarchies learn deeper patterns and see more complex analogies.

The point of these musings is to illustrate that there are many ways brainlike machines could surpass our own abilities, and in dramatic ways. They might think and learn a million times faster than we can, remember vast quantities of detailed information, or see incredibly abstract patterns. They can have senses more sensitive than our own, or senses that are distributed, or senses for very small phenomena. They might think in three, four, or more dimensions. None of these interesting possibilities depend on intelligent machines mimicking or acting like humans, and they don't involve complex robotics.

Now we can see fully how the Turing Test, by equating intelligence with human behavior, limited our vision of what is possible. By first understanding what intelligence is, we can build intelligent machines that are far more valuable than merely replicating human behavior. Our intelligent machines will be amazing tools and will dramatically expand our knowledge of the universe.

How long will it be before any of this comes about? Will we be building intelligent machines in fifty years, twenty years, or five years? There is a saying in the high-tech world that change takes longer than you expect in the short term but occurs faster than you expect in the long term. I have seen this many times. Someone will get up at a conference, announce a new technology, and claim that it will be in every home in four years. The speaker turns out to be wrong. Four years become eight years, and people start thinking it will never happen. Just around that time, when it looks like the whole idea was a dead end, it starts taking off and becomes a big sensation. Something similar will likely happen in the intelligent machine business. Progress will seem slow at first, and then take off rapidly.

At neuroscience conferences, I like to go around the room and ask everyone to state his or her opinion on how long it will be before we have a working theory of cortex. A few people—fewer than 5 percent—say "never" or "we already have one" (surprising answers given what they do for a living). Another 5 percent say five to ten years. Half of the rest say ten to fifty years, or "within my lifetime." The remaining people say fifty to two hundred years, or "not within my lifetime." I side with the optimists. We have been living in the "slow" period for decades, so to many people it seems that progress in theoretical neuroscience and intelligent machines has stalled completely. Judging by the progress in the past thirty years it is natural to assume we are nowhere near an answer. But I believe we are at the turning point and the field is about to take off.

It is possible to accelerate the future, to move the turning point closer to the present. One of the goals of this book is to convince you that, with the correct theoretical framework, we can make rapid progress in understanding the cortex—that with the memory-prediction framework as our guide, we can decipher the details of how the brain works and how we think. This is the knowledge we need to build intelligent machines. If this is the right model, progress can proceed rapidly.

So although I am hesitant to predict when the age of intelligent machines will be a reality, I think that if enough people apply themselves to solving this problem today, we may be able to create useful prototypes and cortical simulations within just a few years. Within ten years, I hope, intelligent machines will be one of the hottest areas of technology and science. I am reluctant to be more specific than this, because I know how easy it is to underestimate the time it takes to make something important happen. So why am I so optimistic about the speed of progress in understanding the brain and building intelligent machines? My confidence stems mostly from the long time I have already spent working on the problem of intelligence. When I first fell in love with brains, in 1979, I felt solving the puzzle of intelligence was something that could be achieved in my lifetime. Over the years, I have carefully observed the decline of AI, the rise and fall of neural networks, and the Decade of the Brain in the 1990s. I have seen how attitudes toward theoretical biology, and theoretical neuroscience in particular, have evolved. I have seen how the ideas of prediction, hierarchical representation, and time have crept into the language of neuroscience. I have seen the progress in my own understanding and that of my colleagues. I became excited about the role of prediction eighteen years ago and have in some ways been testing it ever since. Because I have been immersed in the neuroscience and computer fields for over two decades, perhaps my brain has built a high-level model of how technological and scientific change occurs, and that model predicts rapid progress. Now is the turning point.

EPILOGUE

The astronomer Carl Sagan used to say that understanding something does not diminish its wonder and mystery. Many people fear that scientific understanding entails a trade-off with wonder, as if knowledge leeches the flavor and color out of life. But Sagan was right. The truth is that, with understanding, we become more comfortable with our role in the universe and simultaneously the universe becomes even more colorful and mysterious. Being a tiny speck in an infinite cosmos, alive, aware, intelligent, and creative, is far more interesting than living on a flat, limited Earth at the center of a small universe. Understanding how our brains work does not diminish the wonder and mystery of the universe, our lives, or our future. Our amazement will only deepen as we apply this knowledge to understanding ourselves, building intelligent machines, and then acquiring more knowledge.

Therefore the quest to understand the brain and build intelligent machines is a worthy endeavor and a logical next step for humanity.

With this book, I hope to entice young engineers and scientists to study the cortex, adopt the memory-prediction framework, and build intelligent machines. At its height, artificial intelligence was a big movement. It had journals, degree programs, books, business plans, and entrepreneurs. Neural networks similarly created tremendous excitement as the field blossomed in the 1980s. But the scientific frameworks underlying AI and neural networks were not the right ones to use in building intelligent machines.

I am suggesting we now have a new, more promising path to follow. If you are in high school or college and this book motivates you to work on this technology—to build the first truly intelligent machines, to help start an industry—I encourage you to do so. Make it happen. One of the keys to entrepreneurial success is that you must jump headfirst into a new field before it is 100 percent clear you can be successful. Timing is important. If you jump too early, you struggle. If you wait until the uncertainty lifts, it's too late. I strongly believe that now is the time to start designing and building cortexlike memory systems.

We are already making significant progress. At the Redwood Neuroscience Institute, our application of this theory, led by Dileep George, is helping to solve problems in computer vision that have eluded researchers for decades. The results are so promising that we launched a new business, Numenta, with the mission of developing far-reaching technologies. (If you are interested in following our work, visit www.Numenta.com.)

This field will be immensely important both scientifically and commercially. The Intels and Microsofts of a new industry built on hierarchical memories will be started sometime within the next ten years. Undertaking an endeavor on this scale can be financially risky or intellectually demanding, but it is always worth trying. I hope you will join me, along with others who take up the challenge, to create one of the greatest technologies the world has ever seen.

APPENDIX: TESTABLE PREDICTIONS

Every theory should lead to testable predictions, for experimental testing is the only sure way to determine the validity of a new idea. Fortunately the memory-prediction framework is grounded in biology and leads to several specific and novel predictions that can be tested. In this appendix I list predictions that can falsify and/or support the proposals made in this book. This material is somewhat more advanced than the material in chapter 6 and is definitely not required to understand the rest of the book. Several of the predictions can only be performed on awake animals or awake human subjects because the tests involve expectation and prediction of the onset of a stimulus. The predictions are not ranked in terms of importance.

Prediction 1

We should find cells in all areas of cortex, including primary sensory cortex, that show enhanced activity in anticipation of a sensory event, as opposed to in reaction to a sensory event.

For example, Tony Zador's lab at Cold Spring Harbor

Laboratory has found cells in rat primary auditory cortex that fire precisely when the rat expects to hear a sound even when there is no sound (private correspondence). This should be a general property of cortex. We should find similar anticipatory activity in visual cortex and somatosensory cortex. Cells that fire in anticipation of a sensory input are the definition of prediction, a basic premise of the memory-prediction framework.

Prediction 2

The more spatially specific a prediction can be, the closer to primary sensory cortex we should find cells that become active in anticipation of an event.

If a monkey were trained on sequences of visual patterns such that it could anticipate a particular visual pattern at a precise time, we should find cells that show enhanced activity precisely when the anticipated pattern is expected (a restatement of prediction 1). If the monkey learned to expect to see a face but it didn't know exactly what face or how the face would appear, then we should expect to find anticipatory cells in face recognition areas but not lower visual areas. However, if the monkey fixates on a target and has learned to expect a particular pattern at a precise location in its visual field then we should find anticipatory cells in V1 or close to V1. Activity representing prediction flows down the cortical hierarchy as far as it can, depending on the specificity of the prediction. Sometimes it can go all the way to primary sensory areas and other times it stops in higher regions. Similar results should exist in other sensory modalities.

Prediction 3

Cells that exhibit enhanced activity in anticipation of sensory input should be preferentially located in cortical layers 2, 3, and 6 and the prediction should stop moving down the hierarchy in layers 2 and 3.

Predictions that travel down the cortical hierarchy do so via cells in layers 2 and 3, which then project to layer 6. These layer 6 cells project broadly across layer 1 in the region below in the hierarchy, activating another set of layer 2 and layer 3 cells and so on. Therefore cells in these layers (2, 3, and 6) are where we should find anticipatory activity. Recall that active cells in layers 2 and 3 represent a set of possible active columns; they are possible predictions. Active cells in layer 6 represent a smaller number of columns; these are the specific predictions from a region of cortex. As a prediction travels down the hierarchy the activity will eventually stop in layers 2 and 3. For example, say a rat has learned to anticipate one of two different audio tones. Based on an external cue, the rat knows when it will hear one of these two tones but it can't predict which tone. In this scenario we should expect to see anticipatory activity in layers 2 or 3, in columns that represent both tones. There should not be activity in layer 6 of the same region because the animal can't predict which specific tone will be heard. If on another trial the animal can predict the exact tone it will hear, then we should see activity in layer 6, in columns that respond to that specific tone.

We can't completely rule out the possibility of finding anticipatory cells in layers 4 and 5. For example, it is likely there are several classes of cells in these layers with unknown function. Therefore, this prediction is relatively weak, but I still feel it is worth mentioning.

Prediction 4

One class of cells in layers 2 and 3 should preferentially receive input from layer 6 cells in higher cortical regions.

Part of the memory-prediction model is that learned sequences of patterns that occur together develop a temporally constant invariant representation, what I call a "name." I propose that this name is a set of cells in layers 2 or 3 across a region of cortex in different columns. The set of cells remains active as long as events that are members of the sequence are occurring (e.g., a set of cells that remains active as long as any note in a melody is being heard).

This set of cells representing the name of the sequence is made active via feedback from layer 6 cells in higher regions of cortex. I suggest these name cells are layer 2 cells because of their proximity to layer 1. But it could be any class of cell within layers 2 and 3, which have dendrites in layer 1. For the naming system to work, the apical dendrites of these name cells must form synapses preferentially with axons in layer 1 that originated in layer 6 of higher regions. They should avoid forming synapses with axons in layer 1 that originated in the thalamus. Thus the theory suggests we should find a class of cells, within layers 2 and 3, with apical dendrites in layer 1, that have a strong preference for forming synapses with axons from cells in layer 6 in the region above. Other cells with layer 1 synapses should not have this preference. This is a strong and, as far as I know, novel prediction.

A corollary prediction is we should find another class of cells in layers 2 or 3 whose apical dendrites form synapses preferentially with axons originating in nonspecific regions of the thalamus. These cells predict next items in a sequence.

Prediction 5

A set of "name" cells described in prediction 4 should remain active during learned sequences.

A set of cells that stays active during a learned sequence is the definition of a "name" for a predictable sequence. Therefore we should find sets of cells that remain active even as the activity of cells in the rest of a column (cells in layers 4, 5, and 6) is changing. Unfortunately we can't say what the activity of the name cells will look like. For example, the constant activity of a name pattern could be as simple as a single spike in unison across the set of name cells. Therefore, this group of active cells might be hard to detect.

Prediction 6

Another class of cells in layers 2 or 3 (different from the name cells referred to in predictions 4 and 5) should be active in response to an unanticipated input, but should be inactive in response to an anticipated input.

The idea behind this prediction is that unanticipated events must be passed up the cortical hierarchy, but when an event is anticipated we don't want to pass it up the hierarchy precisely because it was predicted locally. Therefore there should be a class of cells in layers 2 and 3, different from the name class described in predictions 4 and 5, that shows activity when an unanticipated event occurs, but doesn't show activity if the event was anticipated. The axons of these cells should project to higher regions of cortex. I propose one mechanism to achieve this change in activity. Such a cell could be inhibited via an interneuron activated by a name cell, but at this point there is no way to make a solid prediction of the mechanism. All we can say is that some cells should exhibit this differential activity. This is another strong and, as far as I know, novel prediction.

Prediction 7

Related to prediction 6, unanticipated events should propagate up the hierarchy. The more novel the event the higher the unanticipated input should flow. Completely novel events should reach the hippocampus.

Heavily learned patterns are predicted lower in the hierarchy, and, conversely, the more novel an input, the higher it should propagate up the hierarchy. It should be possible to design an experiment to capture this difference. For example, a human could listen to an unfamiliar but simple melody. If the subject hears a note that, although unexpected, is consistent with the style of music, the unexpected note should cause changes of activity in auditory cortex, up to some level in the cortical hierarchy. However, if instead of hearing a note consistent with the style of music the subject heard a complete nonsense sound, such as a crash, we would expect changes of activity from this sound to travel higher up the cortical hierarchy. The results should switch if the subject was expecting to hear the crash but instead heard the note. It might be possible to test this prediction with fMRI on human subjects.

Prediction 8

Sudden understanding should result in a precise cascading of predictive activity that flows down the cortical hierarchy.

The "aha" moment when a puzzling sensory pattern is finally understood—such as recognizing the dalmatian dog in figure 12—begins when a region of cortex attempts a new memory match of its input. If the match fits the local region, predictions are passed down the cortical hierarchy in quick succession to all lower regions. If this is a correct interpretation of the stimulus, then each region of the hierarchy will settle on a correct prediction in rapid succession. The same effect should occur while viewing an image with two interpretations, such as a silhouette of a vase that can look like two faces or a Necker cube (an image of a cube that alternately appears in two different orientations). Every time the percept of such an image changes we should see a propagation of new predictions flow down the hierarchy. At the lowest levels, say V1, a column representing a line segment of the image should stay active in either perception of the image (assuming the eyes haven't moved). However, we might see some cells in that column swap active states. That is, the same low-level feature exists in each image but different cells within a column may be active in the different interpretations. The main point is that we should see a propagation of predictions flow down the cortical hierarchy when a high-level percept changes.

A similar propagation of prediction should occur with each saccade over a learned visual object.

Prediction 9

The memory-prediction framework requires that pyramidal neurons can detect precise coincidences of synaptic input on thin dendrites.

For many years it was thought that neurons might be simple integrators, summing the inputs from all their synapses to determine if the neuron should spike. Within neuroscience today, there is much uncertainty about how neurons behave. Some people still hold to the idea of neurons being simple integrators, and many neural network models are built with neurons that work this way. There are also many neuron models that assume a neuron behaves as if each dendritic section operates independently. The memory-prediction model requires that neurons be able to detect the coincidence of only a few active synapses in a narrow window of time. The model could work with even a single potentiated synapse being sufficient to cause a cell to fire, but more likely it would be two or more active synapses in proximity on a thin dendrite. Thus a neuron with thousands of synapses can learn to fire on many different precise and separate input patterns. This is not a new idea, and there is evidence to support it. It is, however, a radical departure from the standard model used for many years. If it were shown that neurons don't fire on precise and sparse input patterns, it would be difficult to keep the memory-prediction theory intact. Synapses on thick dendrites on or near the cell body need not work this way, only the many synapses on thin dendrites.

Prediction 10

Representations move down the hierarchy with training.

I argue that through repeated training, the cortex would relearn sequences in hierarchically lower regions of cortex. This follows naturally from how the memory of sequences of patterns would change the input pattern passed to the next higher regions of cortex. There are a couple of consequences of this process. One is we should find cells that respond to a complex stimulus lower in the cortex after extensive training and higher in the cortex after minimal training. In a human, for example, I would expect to find cells that respond to printed letters in a region such as IT after training to recognize individual letters. But, after learning to read entire words, I would expect to find cells that respond to letters in different parts of V4 in addition to IT. Similar results should be attainable with other species, other regions, and other stimuli. Another consequence of this learning process is that where recall occurs and where errors are detected should move. That is, sensations of highly learned patterns should propagate less distance up the hierarchy. This might be detectable via imaging techniques. We should also be able to detect a change in reaction times to certain stimuli because inputs will not have to travel as far in the cortex to be recognized and recalled.

Prediction 11

Invariant representations should be found in all cortical areas.

It is well known that cells exist that respond to highly selective inputs invariant to many details. Cells that respond to faces, hands, Bill Clinton, etc., have been observed. The memory-prediction model predicts that all regions of cortex should form invariant representations. The invariant representations should reflect all sensory modalities below a region of cortex. For example, if I had a Bill Clinton cell in visual cortex it would fire whenever I see Bill Clinton. If I had a Bill Clinton cell in auditory cortex, it would fire whenever I hear the name "Bill Clinton." I would then expect to find cells in association areas that receive both visual and auditory input that respond to either the sight or the spoken name of Bill Clinton. We should find invariant representations in all sensory modalities and even motor cortex. In motor cortex, cells would represent complex motor sequences. The higher up the motor hierarchy, the more complex and invariant the representation should be. (Recent studies appear to have found cells that activate complex hand-to-mouth movements in monkeys.) These are not novel predictions. Most researchers believe in the general idea that invariant representations are formed in many locations throughout the cortex. However, even though I discussed this as a fact, it has not been demonstrated everywhere. The memory-prediction model predicts we will see such cells in every region of cortex.

The preceding predictions are some of the ways the model in this book can be tested. I am sure there are others. However, it isn't possible to prove a theory is correct. It is only possible to prove a theory is incorrect. So even if all the predictions I listed above were shown to be true, that wouldn't be proof that the memory-prediction hypothesis is correct, but it would be strong evidence in support of the theory. The flip side is also true. If some of the predictions above turn out not to be true, it wouldn't necessarily invalidate the entire thesis. For some of the predictions there are alternate ways the needed behavior could be achieved. For example, there are other ways names of sequences could be created. This appendix is only intended to show that the model leads to several predictions and, therefore, can be tested. Designing experiments is challenging work and would need much more discussion than is appropriate for this book. It would also be great if we could find ways of testing this theory with imaging techniques such as fMRI. There are many imaging laboratories and these experiments can be performed relatively quickly when compared to recording directly from cells.

BIBLIOGRAPHY

Most science books and journal articles have lengthy bibliographies, which serve as much to catalog the contributions of others as to assist the reader. Given that this book is intended for a variety of readers, including those with no previous neuroscience knowledge, I have avoided writing the book in an academic style. Similarly, this bibliography is designed primarily to assist the nonexpert reader who wants to learn more. I don't list all relevant published research, nor do I attempt to credit all the individuals who have made the fundamental discoveries in this field. Instead I list selected items that I believe would be good materials for an interested reader to learn more about brains. I also include several items that I found useful but are mostly for the specialist. You can find in-depth discussions on many of these topics on the World Wide Web. More bibliographic material can be found on this book's Web site, www.OnIntelligence.org.

Unfortunately, you will find only a few references to overall theories of the brain because, as I wrote in the prologue, not a lot has been written on this topic, and even less on the specific proposals outlined in this book.

History of AI and Neural Networks

Baumgartner, Peter, and Sabine Payr, eds. *Speaking Minds: Interviews with Twenty Eminent Cognitive Scientists* (Princeton, N. J.: Princeton University Press, 1995).

This book contains interesting interviews with many of the leading thinkers in AI, neural networks, and cognitive science. It is an easy and enjoyable synopsis of the recent history and spirit of thinking on intelligence.

Dreyfus, Hubert L. *What Computers Still Can't Do: A Critique of Artificial Reason* (Cambridge, Mass.: MIT Press, 1992).

A harsh critique of AI originally published under the title *What Computers Can't Do* and reissued years later with the revised title. It is an in-depth history of AI written by one of its strongest critics.

Anderson, James A., and Edward Rosenfeld, eds. N*eurocomputing, Foundations of Research* (Cambridge, Mass.: MIT Press, 1988).

This large book is an annotated collection of important papers in neutral network and brain theory spanning the years 1890 to 1987, presented in chronological order. It contains papers by W. S. McCulloch and W. Pitts, Donald Hebb, Steve Grossberg, and many others, with an introduction to each paper by the editors. It is an easy way to read many of the important historical papers in this field.

Searle, J. R. "Minds, Brains, and Programs," *The Behavioral and Brain Sciences*, vol. 3 (1980): pp. 417–24.

Presents the famous "Chinese Room" argument against computation as a model for the mind. You can find many descriptions and discussions of Searle's thought experiment on the World Wide Web.

Turing, A. M. "Computing Machinery and Intelligence," *Mind*, vol. 59 (1950): pp. 433–60.

Presents the famous "Turing Test" for detecting the presence of intelligence. Again, many references and discussions on the Turing Test can be found on the World Wide Web.

Palm, Günther. *Neural Assemblies: An Alternative Approach to Artificial Intelligence* (New York: Springer Verlag, 1982).

To understand how the cortex works and how it stores sequences of patterns, it helps to be familiar with auto-associative memories. And although much has been written on auto-associative memories, I have not found any printed sources that present an easily digested summary of what I consider important. Palm is one of the pioneers in this field. This book of his is hard to obtain and not that easy to read, but it covers the basics of auto-associative memories including sequence memory.

Neocortex and General Neuroscience

The following books are recommended for those who want to learn more about neurobiology and the neocortex.

Crick, Francis H. C. "Thinking about the Brain," *Scientific American*, vol. 241 (September 1979): pp. 181–88. Also available in *The Brain: A* Scientific American *Book* (San Francisco: W. H. Freeman, 1979).

This is the paper that got me interested in brains. Although it is twenty-five years old, I still find this paper by Francis Crick inspiring.

Koch, Christof. *Quest for Consciousness: A Neurobiological Approach* (Denver, Colo.: Roberts and Co., 2004).

There are several general-interest brain books published every year. This one by Christof Koch is about consciousness but it covers most of the relevant topics on brains, neuroanatomy, neurophysiology, and consciousness. If you want a basic introduction to neurobiology and brain science in a single readable book, this would be a good place to start.

Mountcastle, Vernon B. *Perceptual Neuroscience: The Cerebral Cortex* (Cambridge, Mass.: Harvard University Press, 1998).

A great book dedicated to everything and anything about the neocortex. It is well written, has a clean layout, and, although technical, I find it a joy to read. It is one of the best introductions to the neocortex.

Kandel, Eric R., James H. Schwartz, Thomas M. Jessell, eds. *Principles of Neural Science*, 4th ed. (New York: McGraw-Hill, 2000).

This is a one-volume encyclopedia of all things neural. This big book is not for bedtime reading but it is a good reference book to have. It provides detailed introductions to all parts of the nervous system, including neurons, sensory organs, and neurotransmitters.

Shepherd, Gordon M., ed. *The Synaptic Organization of the Brain*, 5th ed. (New York: Oxford University Press, 2004).

This book has been helpful to me, although I preferred the earlier editions, which had a single author. It is a technical resource on all parts of the brain, especially synapses. I use it as a reference.

Koch, Christof, and Joel L. Davis, eds. *Large-scale Neuronal Theories of the Brain* (Cambridge, Mass.: MIT Press, 1994).

There is very little written on overall theories of the brain. This book is a compilation of papers on this very topic, although most of the papers in this volume fall short of the goal suggested by the title. This book gives an overview of the varied approaches people are taking to understanding how the overall brain works. You can find bits and pieces of the memory-prediction framework throughout the book.

Braitenberg, Valentino, and Almut Schüz. *Cortex: Statistics and Geometry of Neuronal Connectivity*, 2nd ed. (New York: Springer Verlag, 1998).

This book describes the statistical properties of the mouse brain. I know that doesn't sound very exciting but it is a refreshing and useful book. It tells the story of the cortex in numbers.

Specific Neuroscience Articles

The following articles are the original sources for some of the important concepts described in this book. Most of these can only be found in a university library or online.

Mountcastle, Vernon B. "An Organizing Principle for Cerebral Function: The Unit Model and the Distributed System," in Gerald M. Edelman and Vernon B. Mountcastle, eds., *The Mindful Brain* (Cambridge, Mass.: MIT Press, 1978).

This paper is where I first read Mountcastle's proposal for how the entire neocortex works on a common principle. Mountcastle also proposes that a cortical column is the basic unit of computation. These ideas are both the premise and the inspiration for the theory proposed in this book.

Creutzfeldt, Otto D. "Generality of the Functional Structure of the Neocortex," *Naturwissenschaften,* vol. 64 (1977): pp. 507–17.

After I finished writing *On Intelligence,* I became aware of this paper, which, like Mountcastle's, argues for a common cortical algorithm. It was published slightly before Mountcastle's and is a nice complement to his.

Felleman, D. J., and D. C. Van Essen. "Distributed Hierarchical Processing in the Primate Cerebral Cortex," *Cerebral Cortex,* vol. 1 (January/February 1991): pp. 1–47.

This is the now-classic paper describing the hierarchical organization of the visual cortex. The memory-prediction framework is built on the assumption that not only the visual system but the entire neocortex has a hierarchical structure.

Sherman, S. M., and R. W. Guillery. "The Role of the Thalamus in the Flow of Information to the Cortex," *Philosophical Transactions of the Royal Society of London*, vol. 357, no. 1428 (2002): pp. 1695–708.

Provides an overview of thalamic organization and lays out the Sherman-Guillery hypothesis in which the thalamus serves to gate information flow between cortical areas. I elaborate on this idea in chapter 6 in the section titled "An Alternate Path up the Hierarchy."

Rao, R. P., and D. H. Ballard. "Predictive Coding in the Visual Cortex: A Functional Interpretation of Some Extra-Classical Receptive-field Effects," *Nature Neuroscience*, vol. 2, no. 1 (1999): pp. 79–87.

I include this paper as an example of recent research that discusses prediction and hierarchies. Rao and Ballard's paper presents a model of feedback in cortical hierarchies, in which neurons in higher areas attempt to predict patterns of activity in lower areas.

Guillery, R. W. "Branching Thalamic Afferents Link Action and Perception," *Journal of Neurophysiology*, vol. 90 (2003): pp. 539–48.

Young, M. P. "The Organization of Neural Systems in the Primate Cerebral Cortex," *Proceedings of the Royal Society: Biological Sciences*, vol. 252 (1993): pp. 13–18.

These two well-written papers provide evidence that motor behavior and sensory perception are intimately related and part of the same process. Guillery argues that all sensory cortical areas play a role in motor behavior and Young shows that motor cortex and somatosensory cortex are so tightly linked that they should be considered one system. I briefly discuss these ideas in chapter 6.

ACKNOWLEDGMENTS

Whenever someone asks me, "What do you do for a living?" I never know what to say. For the truth is I do very little. But I have surrounded myself with people who seem to get a lot done. My contribution is to nudge them every once in a while, and when necessary try to redirect the team down a new path. To the extent that I have had success in my career, I owe it mostly to the hard work and intelligence of my colleagues.

I have had the chance to meet many scientists and almost all of them have taught me something and therefore they have all contributed to the ideas in this book. I thank them all but can mention only a few here. Bruno Olshausen, who has joint positions at the Redwood Neuroscience Institute (RNI) and the University of California at Davis, is a walking encyclopedia of neuroscience. He is constantly pointing out what I don't know and suggesting ways to rectify my ignorance, which is one of the most valuable things a person can do. Bill Softky, also at RNI, was the first person to teach me about reduction in time in the cortical hierarchy and the properties of thin dendrites. Rick Granger at the University of California at Irvine gave me insights into sequence memory and how the thalamus could play a role. Bob Knight at the University of California at Berkeley and Christof Koch at the California Institute of Technology have been instrumental in the formation of the Redwood Neuroscience Institute and many other science matters. All the staff at RNI have challenged me and forced me to refine my ideas; many of the proposals in this book were a direct result of meetings and discussions at RNI. Thank you all.

Donna Dubinsky and Ed Colligan have been my business partners for a dozen years. Through their hard work and assistance, I was able to be an entrepreneur while simultaneously working part-time on brain theory, an unusual arrangement. Donna used to say one of her objectives was to make our businesses successful so I could afford to spend more time on brain theory. This book wouldn't exist if it weren't for Donna and Ed.

I couldn't have written *On Intelligence* without lots of help. Jim Levine, my agent, believed in this book even before I knew what I would write. Don't write a book without an agent like Jim. He introduced me to Sandra Blakeslee, my coauthor. I wanted this book to be readable by a wide audience and Sandy was essential for achieving that. I take all the blame for the difficult sections. Matthew Blakeslee, Sandy's son and also a science writer, provided several examples used in the book and suggested the term *memory-prediction framework*. All the folks at Henry Holt have been great to work with. I would like to especially thank John Sterling, president and publisher of Henry Holt. I met John face-to-face only once, and we talked on the phone only a few times. That is all it took for him to have a huge impact on the structure of this book. He immediately understood the issues I would face proposing a theory of intelligence and then he suggested how the book should be written and positioned.

I would like to thank my daughters, Anne and Kate, for not complaining when their dad spent many weekends at the computer keyboard. And, finally, I want to thank my wife, Janet. Being married to me can't be all that easy. I love her more than brains.

INDEX

A1 (primary auditory area), 46, 55, 115
action potentials/spikes, 56, 110, 162, 163
AI, *see* artificial intelligence (AI)
alien limb syndrome, 200
altered door experiment, 87–89, 119
Alzheimer's disease, 200
analogies, 180, 181, 188
  false, 192–93, 216, 227
  finding, 189–90, 191–92
  intelligent machines, 209
  prediction by, 183, 184–87, 203
animals, intelligence in, 177–80
anterior cortex, 102–3
anticipation, 148, 149
anticipatory cells, 237–41
  see also prediction
artificial intelligence (AI), 3, 4, 5, 9–22, 24, 27, 28, 29, 31, 32, 33, 37, 38, 40, 65–66, 88, 90, 136, 209, 214–15, 220, 233, 235
association areas/regions, 46, 47, 60–61, 114, 117, 119, 120, 122, 123, 125, 130
  in memory recall, 71, 73, 74
attention, mechanism, 171–74
attention when something is wrong, 86, 89, 93, 95–96
auditory cortex, 63, 81, 115, 117, 131
auto-associative memory, 29–31, 70, 73–75, 76, 80, 105, 164
  used to store sequences of patterns, 144, 146

basal ganglia, 41, 168, 169
Bayes, Thomas, 90
Bayesian networks, 91
behavior, 18, 38, 104, 120
  and intelligence, 6, 29, 32–33, 97, 100, 101–5, 231
  means of exploiting structure of world, 178, 179–80
  neocortex in, 69, 104, 120, 130–31, 144–46, 157–58
  perception and, 157
  prediction and, 89, 97–98
behaviorism, 16
Beksey, Georg von, 58
Blocks World, 17
Brahe, Tycho, 193
brain(s), 1–2, 3, 9–10, 21–22, 32, 52–53, 85
  architecture of, 25, 26, 40
  capacity of, 224–26
  complexity of, 174
  as computer, 65–68

brain(s) (*cont.*)
computer different from, 2, 5, 12, 40, 65–69
computer simulating, 21
criteria for understanding, 25
evolution of, 98–99
flexibility of, 53–54, 62, 228
as machine, 65–66
models of, 25, 233
and patterns, 62–64, 199–200
physical variation, 188–89
predictions, 6, 86–87, 88–89, 90, 96, 97
storing sequences of sequences, 129
structure of, 23–24, 26, 182
time-related constraints, 224
training, 226
Broca's area, 44, 50
building intelligent machines, 1, 2, 4–5, 7, 10, 11, 13, 29, 35–36, 37, 41, 62–63, 89, 176, 203–13, 234, 235
challenges in, 210–13
creativity and, 183
ethical/moral issues, 213–17
recipe for, 209
Turing and, 14
why build, 217–22

capacity (intelligent machines), 210–11, 224–26
central sulcus, 103
cerebellum, 41, 168, 169
Chinese Room (thought experiment), 18–20, 38, 86, 105, 218
cochlea, 58–59, 81
coincidences, 163, 164
columns, cortical, *see* cortical columns
computation/computing, 13–15, 68–69
computer(s), 9, 13, 28, 37–38, 90
brain as, 65–68
intelligence, 5, 14–15, 16, 20, 21
pattern recognition, 190–91
speech recognition, 218–19
connectionists, 24, 29, 37, 39
connectivity, 109–10
intelligent machines, 210, 211–12
in neural networks, 24, 45
consciousness, 6, 193–200
cortex, *see* neocortex/cortex
cortical algorithm, 51, 52, 55, 62, 63, 64, 80, 96, 99, 101, 115, 122, 124, 125, 127, 204, 215, 216, 221, 224, 226
finding patterns, 228
cortical columns, 139–42, 144, 146, 147, 148–52, 154–55, 156, 159, 163, 164
auto-associative memory formed out of, 144
basic unit of prediction, 141
cortical hierarchy, 56, 105, 137, 164–65
alternate path up, 171–74
alternate view of, 123–25, 123*f*
feedback, 162–63
higher regions, 149, 150, 153
hippocampus at top of, 170
information flowing up/down, 46–47, 141–42, 142*f*, 147, 158–61, 162, 167, 171–74
invariant representations, 121, 122
memories, 166, 171, 188
predictions moving down, 238–39
reciprocal connections, 161–62
representations in, 243–44
sequences in, 130–33, 158
and unanticipated events, 241–42
cortical layers, 42
anticipatory cells in, 238–39
learning and memory in, 164
in cortical region, 147–66, 172
cortical region/areas, 50–51, 62, 113, 114, 120–25, 137–41, 146–47, 157, 158–61, 170–71
anticipatory activity in, 237–38
building sequences, 165–66
classifying inputs, 147–48
communicating with each other, 144–47
comparing expected/observed columns, 156
connections between, 114, 172, 211
and creativity, 183
imagination in, 200–1
invariant representations in, 124–25, 244–45
layers and columns in, 138*f*
making predictions, 147, 153–56
"name cells," 239–40
sequence of patterns in, 128, 129–37, 147, 148–49, 150–53

size of, 138–39, 141
subregions, 122, 123, 124–25
cortical sheet, 42, 45, 47, 55, 99, 110, 168, 180, 211
creativity, 6, 89, 183–93, 213
differences in, 187–89
training for, 189–92
Crick, Francis, 10, 13, 43
culture, and world model, 202, 203

Darwin, Charles, 33, 50
Decade of the Brain, 31, 233
declarative memories, 196–97
Deep Blue, 17, 18, 20–21
dendrites, 48, 162, 163, 164
dinner napkin analogy, 42, 109, 122
DNA and genetic memory, 178, 182
dolphins, 103–4

Einstein, Albert, 49, 189, 225
Eliza, 16–17
emotion, 208, 216, 230
Erdös, Paul, 8
Essen, David van, 45
ethical/moral issues
in building intelligent macines, 213–17
evolution, 33, 37, 38, 39, 41, 170, 208–9, 225
beginning of intelligence in, 177, 178, 179–80
of neocortex, 42, 99, 101, 103, 104, 180
expertise, 166–68, 188, 227
expert systems, 17–18

face cell, 113
face recognition, 77, 81–82, 111, 113, 121
feedback, 25, 26, 31, 32, 45, 90, 113, 161–64
in auto-associative memories, 29–30
delayed, 146, 147
folded, 156, 201
in sensory regions, 46
feedforward-feedback process, 113–14, 153
Felleman, Daniel, 45
filling in, 93–94, 99
fixation(s), 57, 58, 94–95, 110, 111, 112, 116, 128, 130, 132
forms platonic, 78–79, 82
Fried, Itzhak, 108
functional imaging, 31–32, 52–53, 245
functional areas/regions, 44, 45, 52–54, 109–10
functionalism, 36–38

genius, 168, 183
goal-oriented behavior, 158
Graffiti, 28, 191–92
Grid Systems, 21, 22, 175
Grossberg, Stephen, 90, 156, 201
Guillery, Ray, 172

hallucination, 63–64
hearing, 51, 120, 156–57, 158, 198
invariant representations, 114, 114*f*, 115, 116
spatial/temporal patterns in, 58–59
Hebb, Donald O., 164
Hebbian learning, 48, 164
hierarchical memory system, 208–10, 221–22, 235
hierarchical structure of neocortex, 6, 25, 70, 107, 109, 125–27, 129, 176, 180
see also cortical hierarchy
hierarchy, 44–47, 55
of intelligent systems, 231
learning in, 165–68
patterns stored in, 70
hippocampus, 42, 168–71, 241
Hoff, Ted, 11

imagination, 40, 156, 177, 200–1
information flows, 45, 54, 69, 113
in brain, 55–57
sensory hierarchies, 117–20, 118*f*, 124
within six-layered cortex, 52
up and down cortical hierarchy, 46–47, 141–42, 147, 158–61, 162, 167, 171–74
inhibitory cells, 147–48, 151
input, flow of over time, 116
input neurons, 25, 26
input-output fallacy, 31–32, 37
intelligence, 1, 3, 11, 225–26
animal, 177–80
and behavior, 6, 29, 32–33, 97, 231

intelligence (*cont.*)
and emotion, 216
epochs of, 182
existence proof of, 14
future of, 205–33
human, 180
measure of, 210
memory and prediction in, 99–105
real intelligence, 4–5, 13
intelligence tests, 96
intelligent machines, 5, 32, 33, 39, 41, 55, 90, 206, 232
capabilities of, 222–33
capacity, 210–11, 224–26
connectivity, 210, 211–12
possibility of building, 203–13
progress toward, 232–33
replicability, 226–27
as threat, 214–17
uses of, 217–22
see also building intelligent machines
invariance, 79, 131
invariant memory, 105, 203, 228
in predictions, 183–84
invariant prediction, 153–56, 155*f*
invariant representations, 69, 75–82, 83–84, 107, 108, 109–16, 153, 180, 203
converting into predictions, 143–44
formed in cortical hierarchy, 121, 122, 243–44
forming, in hearing, vision, and touch, 114*f*
found in all cortical areas, 124–25, 134, 244–45
motor behavior represented in hierarchy of, 157–58
sequences and, 128, 130, 137, 138
IT, 45, 110, 112–13, 121, 122, 125, 129, 130, 157, 167, 172, 189

Joy, Bill, 214

Kasparov, Gary, 17
Keller, Helen, 62, 64
Kepler, Johannes, 192–93
kludge, 37
Koch, Christof, 108, 193
Kreiman, Gabriel, 108
Kurzweil, Ray, 214

language, 12, 16, 18, 23, 40, 64, 97, 102, 104, 125, 181–82
invention of, 182, 183
machines understanding, 218, 219, 220
layers, cortical, *see* cortical layers
learned sequences
name cells active during, 240
learning, 6, 31, 78, 164–68, 178, 179–80, 188, 226–27
essence of, 136
hippocampus in, 169
Llinas, Rodolfo, 90
logic gates, 15

M1, 46
McCulloch, Warren, 15–16
Mackay, D. M., 90
memory, 64, 65–84, 86, 93, 126, 178–79
in consciousness, 199
erasing, 196–97
hierarchical, 166–67
hippocampus and, 168, 169
and intelligence, 99–105, 182
in problem solving, 68–69
memory chip, 206, 211
memory-prediction framework, 5, 104, 105, 107–9, 120, 140–41, 163–64, 177, 180–83, 185, 187–89, 233, 235, 242–43
testable predictions, 237–45
memory recall, 71–72, 73, 82
memory system, 174, 176
adding to sensory input, 99–100, 101, 102, 104
hierarchical, 208–10, 221–22, 235
memory systems, brainlike, 206, 209–11, 212, 215, 216, 224–26, 227, 235
scaling beyond biological brain, 223–32
uses for, 217–18, 219–20
microcolumn, 140
microprocessor, 9, 10, 11, 206, 218, 222
mind, 36, 199, 200, 204, 208
same as brain, 43
as software, 37–38
Moore, Gordon, 10, 11, 217
Moore's Law, 217, 218
motor behavior, 104, 157–58
motor control, 51, 168–69

motor cortex, 103, 120, 131, 144–45, 157
invariant representations in, 80, 137
language and, 181
motor system, 46, 168
Mountcastle, Vernon, 50–52, 53, 55, 80, 96, 120, 121, 122, 140–41
Mumford, David, 90
music, 40, 80–81, 115, 125, 167, 187
myelin/white matter, 78, 144, 146, 211, 212

name cells, 150–53, 151*f*, 239–40
active during learned sequences, 240
"names," 129–30, 132, 136, 151, 159
cortical region forming, 147, 149, 150–53, 151*f*
neocortex/cortex, 3, 13, 25, 29, 38, 39, 40–42, 43–44, 49–52, 74–75, 99–105, 198, 235
architecture of, 55, 107, 120–21, 126–27
common algorithm/function, 51, 52, *see also* cortical algorithm
and consciousness, 196, 199–200
evolution of, 42, 99, 101, 103, 104, 180
flexibility of, 53–54, 55, 61–62, 136
formation of, 140
hierarchical structure of, 6, 25, 70, 107, 109, 125–27, 129, 176, 180, *see also* cortical hierarchy
hierarchies, 44–47
higher regions of, 165, 166
hippocampus and, 170
how it learns, 164–68
information flowing into, 45, 69, 170–71
and intelligence, 6, 98–105
large size, 100, 101, 103, 104, 182
motor control, 168–69
plasticity, 54
prediction(s), 82–84, 88–89, 92–97, 113
six-layered form, 42–43, 47, 48, 51, 52, 56, 69, 107, 109, 139–47 *see also* cortical layers
size of, and intelligence, 180, 181, 183
survival value of, 97
thalamus and, 25
using stored memories, 69–70
neocortical memory, 70–84, 100–1
invariant representations, 75–79
neocortical pyramid, 170, 171
Nestor, 28
NetTalk, 27
neural networks, 5, 11, 23–39, 40, 233, 235
neurons, 42–43, 47–48, 77, 85, 86, 89, 162, 163, 164, 176, 204, 211
coincidence detectors on thin dendrites, 242–43
connections, 24, 25–26, 29–30, 39
evolution of, 179–80
as logic gates, 15–16
and memory, 68, 69, 73, 92
speed of, 66–67, 88, 223
nurture, 187–88, 202–3

object recognition, 113, 116, 136–37
"old"/"primitive" brain, 98, 99–100, 102, 103, 180
emotional systems of, 208, 216
motor system of, 104, 157
Olshausen, Bruno, 170
one-hundred-step rule, 66–68, 78
optic nerve, 55, 56, 110, 199

PalmPilot, 1, 28
pattern classification, 134, 147–48, 136–37, 165, 166
patterns, 56–64, 65, 74, 108, 110, 115–16, 178, 181, 192
analogous, 189
in auto-associative memory, 30–31, 70, 73–75, 144, 146, 181
equivalency, 60–61
grouping, 165
intelligence and, 63
intelligent machines finding, 228, 229, 230, 231
learning as sequence of, 31
in neocortex, 70, 120, 125
in neural networks, 25–26
recognition of, 188
stored as memories, 97
stored in hippocampus, 171

patterns (*cont.*)
stored in invariant form, 70
transformed, 76–77
unexpected, 152, 159–60, 174, 186–87, *see also* sequence of patterns
perception(s), 13, 18, 40, 44, 56, 63, 80, 87, 104, 157, 183
Pitts, Walter, 15–16
plasticity, 54, 179–80, 182
Plato, 17, 78
Platonic solids, 192–93
posterior cortex, 102–3
predictability, 128–29
prediction(s), 32, 82–84, 86–87, 88–89, 92–97, 107, 113, 119–20, 136, 138, 147, 148, 153–62, 177, 180, 182, 188, 233, 238–39
and behavior, 97–98, 101
column basic unit of, 141
in consciousness, 199
converting invariant representations into, 143–44
and creativity, 183–84
foundation of intelligence, 84, 89–90, 96, 97, 99–105
in imagination, 200–1
invariant converted to specific, 153–56, 155*f*
multisensory, 117–19
and reality, 202
using for successful reproduction, 178–79, 180
prediction by analogy, 180, 184–87, 203, 205
predictions, testable, 108, 237–45
primary sensory areas, 45–46, 138
anticipatory cells, 237–38
"primitive" brain, *see* "old"/"primitive" brain
probabilistic predictions, 92–93
probability theory, 90
proprioceptive system, 59
Proust, Marcel, 74
pyramidal neurons, 48–49

qualia, 196, 197–99

Rao, Rajesh, 90
reality, 63–64, 128–29, 177, 201–4
Redwood Neuroscience Institute, 3–4, 170
religious beliefs, 203
replicability
intelligent machines, 226–27
retina, 57, 59, 110–11, 112
robotics, 12, 32, 158, 223, 231
robots, 2, 7, 80, 207–8

S1 (primary somatosensory area), 46
saccade(s), 57, 58, 94–95, 110, 111*f*, 112, 128–29, 130, 132, 157–58
savants, 225–26
Searle, John, 18–20, 38, 86, 105, 218
senses, 55, 59, 64, 208–9, 228–29
integrating, 117–20
and intelligence, 62
qualitative difference, 198
sensory cortex, 80–81, 120, 157
sensory information/input, 45–46, 55–56, 61, 89, 109, 169, 173
adding memory system to, 99–100, 101, 102, 104
in hierarchies, 117–120, 118*f*, 123, 124
spatial-temporal patterns, 59
stored/ recalled, 99–100, 101, 102
sensory systems
intelligent machines, 225, 227–32
sequence(s), 127–28, 129, 183–84
forming, 134–35, 165–66
memory of, 166
stored, 84, 144
sequence of patterns, 70–73, 127–28, 129, 130–31, 159
in cortical regions, 128, 129–37, 138
cortical regions learning and recalling, 147, 148–49, 153
storing, 105, 125, 148
sequences of sequences, 129–33
Shakespeare, William, 186
Sherman, Murray, 172
sight, 51, 59, 117, 118, 119, 198
silicon, 10–11, 204, 210–11, 212, 217, 223–24
smart cars, 218, 220–22, 226–27
somatosensory cortex, 46, 56, 72, 115, 116, 117
soul, 78, 199
spatial patterns, 57–59, 63, 65, 81, 110, 164, 199
recalling, 73–74
recognized with brief fixation, 116

speech recognition, 218
stereotypes, 202, 203–4
sulcus/sulci, 44, 189
synapses, 48–49, 66, 140, 141, 162–63, 174, 176
  and columns, 148, 149–50
  memories stored in, 73
  modified by experience, 164
  role in cell firing, 163
synesthesia, 198

temporal patterns, 57–59, 63, 65, 81, 110, 199
  recalling, 73–74
thalamus, 42, 48, 55, 147, 157
  alternate pathway through, 172, 173
  current state and current motor behavior communicated within, 145*f*
  neocortex and, 25
Theory of Forms, 78–79
time, 31, 32, 116, 165, 233
  in brain function, 25, 26
  in touch, 59
  in vision, 58
touch, 55, 56, 117, 118, 119, 120, 198
  invariant representations, 114, 114*f*, 115–16
  spatial-temporal patterns, 59
training
  brain, 226
  for creativity, 189–92
  intelligent machines, 209
  representations move down hierarchy during, 243–44

Turing, Alan, 13–14, 15, 18, 29, 105
Turing Machine, *see* Universal Turing Machines
Turing Test, 14, 16, 41, 208, 231–32

Universal Turing Machines, 14, 15, 18, 38

V1 (primary visual area), 45, 47, 50, 110–12, 113, 120–25, 129, 158, 166, 172, 188, 189
  columns in, 139
  pattern of activity in, 77
  size of, 138
V2 and V4, 45–46, 110, 113, 121, 122, 123, 124, 145, 167, 172
vestibular system, 59
vision, 12, 16, 51, 55, 59, 112, 198
  intelligent machines and, 218, 220
  invariant representations, 114, 114*f*, 116, 124
  prediction in, 94–95
  spatial/temporal patterns in, 57–58
visual cortex, 46, 54, 63, 110, 119, 124, 145, 157, 167
  feedback/feedforward connections in, 113

Weihenmayer, Erik, 61
white matter, 78, 144, 146, 211, 212

Zador, Tony, 237
zombies, 194, 195, 197

ABOUT THE AUTHORS

JEFF HAWKINS is one of the most successful and highly regarded computer architects and entrepreneurs in Silicon Valley. Currently the chief technology officer at palmOne, he founded Palm Computing and Handspring and created the Redwood Neuroscience Institute to promote research on memory and cognition. He is a member of the National Academy of Engineering and serves on the scientific board of Cold Spring Harbor Laboratory. He lives in northern California.

SANDRA BLAKESLEE has been writing about science and medicine for the *New York Times* for over thirty years and is the coauthor of *Phantoms in the Brain* by V. S. Ramachandran and of Judith Wallerstein's bestselling books on psychology and marriage. She lives in Santa Fe, New Mexico.